XIᵗʰ INTERNATIONAL ASTRONAUTICAL CONGRESS
STOCKHOLM 1960

XI. INTERNATIONALER ASTRONAUTISCHER KONGRESS

XIᵉ CONGRÈS INTERNATIONAL D'ASTRONAUTIQUE

PROCEEDINGS

VOL. III

THIRD COLLOQUIUM

ON

THE LAW OF OUTER SPACE

EDITED BY

ANDREW G. HALEY

WASHINGTON D.C., U.S.A.

AND

KURT GRÖNFORS

GOTHENBURG, SWEDEN

PUBLISHED BY

THE ORGANIZING COMMITTEE OF THE CONGRESS

STOCKHOLM 1961

ISBN 978-3-662-37067-4   ISBN 978-3-662-37770-3 (eBook)
DOI 10.1007/978-3-662-37770-3

AB AETATRYCK, Ahlén & Akerlunds Tryckerier, Stockholm 1961

Contents

Opening words 1

Session 1. International Control of Outer Space
Jenks, C. W.: The International Control of Outer Space (Introductory Lecture) 3

Discussion. Contributions by:

Rinck, G. 16
Smirnoff, M. 16
Safavi, H. 17
Fasan, E. 18
Gunnarsen, L. A. 19
Jenks, C. W.: Concluding remarks of Session 1 19

Papers, Session 1.
Cooper, J. C.: International Control of Outer Space. Some Preliminary Problems 21
Hyman, W. A.: Sovereignty over Space 26
Haley, A. G.: Survey of Legal Opinion on Extraterrestrial Jurisdiction 37
Galloway, E.: World Security and the Peaceful Uses of Outer Space 93
Martin, T. E.: International Space Law and Outer Space 102
Kopal, V.: Two Problems of Outer Space Control: The Delimitation of Outer Space,
and The Legal Ground for Outer Space Flights 108
Gross, F.: Thoughts on the Importance and Task of Space Law 113
Smirnoff, M.: The Today Real Possibilities for the Conclusion of an International
Convention on Outer Space 116
Faria, J. E.: Draft to an International Covenant for Outer Space. The Treaty of Ant-
arctica as a Prototype 122
Michael, D. N.: A Research Approach to the Impacts on Society of Peaceful Space
Activities (Abstract) 128

Session 2. Damage to Third Parties on the Surface Caused by Space Vehicles
Pépin, E.: Damage to Third Parties on the Surface Caused by Space Vehicles (Introductory Lecture) 131

Discussion. Contributions by:

Cooper, J. C. 133
Verplaetse, J. G. 134
Beresford, S. M. 135
von Rauchhaupt, F. W. 136
Pépin, E.: Concluding Remarks of Session 2 137

Closing words 137

Papers, Session 2.

Cooper, J. C.: Memorandum of Suggestions for an International Convention on Third
Party Damage Caused by Space Vehicles 141
Verplaetse, J. G.: Conflicts of Air- and Outer Space Law 145
Beresford, S. M.: Principles of Spacecraft Liability 152

Appendix: Statutes of the International Institute of Space Law 159

List of Authors and Contributors

Numbers refer to page

Beresford, Spencer M. (135, 152)

Special Counsel, Committee on Science and
Astronautics, House of Representatives
United States Congress, Washington, D.C.,
USA

Cooper, John Cobb (21, 133, 141)

L.L., M., L.L.D., (Hon.) Professor Emeritus
International Air Law, Mc Gill University,
Montreal, Canada
1, Armour Road, Princeton, N.J., USA

Faria, J. Escobar (122)

L.L.B., Government Lawyer for the State of
São Paulo, Brazil, Member Air Law Division
and Member Space Sciences Division, The
Santos Dumont Foundation, São Paulo, Brazil
Rua General Mena Barreto, 527, São Paulo,
Brazil

Fasan, Ernst (18)

Dr., Rechtsanwalt
Hauptplatz 11, Neunkirchen, Austria

Galloway, Eilene (93)

Special Consultant, Committee on Aeronautical and Space Sciences
United States Senate, Washington, D.C., USA

Gross, Franz (113)

Dr., Rechtsanwalt
Pestalozzistrasse 1, Graz, Austria

Grönfors, Kurt (1, 137)

Professor of Law, Gothenburg's School of Eco-
nomics and Business Administration
Handelshögskolan, Vasagatan 3, Gothenburg,
Sweden

Gunnarsen, Leif (19)

Landsretssagfører, Sekretær i Undervisnings-
ministeriet, København
Ingolfs Alle 24 B, Copenhagen S, Denmark

Haley, Andrew G. (37)

General Counsel, International Astronautical
Federation
General Counsel American Rocket Society
1735 DeSales Street, N.W., Washington 6, D.C.,
USA

Hyman, William A. (26)

Attorney at Law, A.B., L.L.B., D.H.L.
111 Fulton Street, New York 38, N.Y., USA

Jenks, C. Wilfred (3, 19)

L.L.D., Associate of the Institute of International Law
International Labour Office, Geneva, Switzerland

Kopal, Vladimír (108)

Dr., Institute of State and Law, Czechoslovak Academy of Sciences, Praha, Czechoslovakia
Praha 1, Národní 18, Czechoslovakia

Martin, Thos. E. (102)

United States Senator, Member of the Committee on Aeronautical and Space Sciences
United States Senate, Washington, D.C., USA

Michael, Donald N. (128)

Dr., Member Senior Staff Brookings Institution
6012, Onondaga Rd., Washington 16, D.C., USA

Pépin, Eugène (131, 137)

Dr., Former Director of the Institute of Air and Space Law, Mc Gill University, Montreal, Canada
51, Rue de Lévis, Paris 17e, France

von Rauchhaupt, Friedrich W. (136)

Professor, Dr.
Plöck 45-49, Heidelberg, Germany

Rinck, Gerd (16)

Professor, Dr.
Herzberger Landstrasse 26, Göttingen, Germany

Safavi, Hassan (17)

Dr., Legal Adviser of Civil Aviation, Iran
25, Ave. Yekta Zafranieh, Tadrish, Teheran, Iran

Smirnoff, Michel (16, 116)

Dr., Chairman International Institute of Space Law, Member Air Transport Commission of International Chamber of Commerce
Zahumska br. 37/III, Belgrade, Yugoslavia

Verplaetse, Julian G. (134, 145)

Dr., S.J.D. Harward
Heirenthoek, Landegem, Belgium

A selection of papers of the other sessions of the congress are to be found in the following publications:

Main Sessions. Proceedings Vol. I, Springer-Verlag, Vienna.

Small Sounding Rockets Symposium. Proceedings Vol. II, Springer Verlag, Vienna.

Space Medical Symposium and Astrodynamics Colloquium — in forthcoming issues of Astronautik (Journal of the Swedish Interplanetary Society, Stockholm).

Third Colloquium on the Law of Outer Space

Opening Words

Kurt Grönfors

Ladies and Gentlemen:

As Chairman of the Swedish Committee organizing this Colloquium I have the great honour and pleasure of greeting you all most heartily welcome. Indeed, I am very happy to see present in this room at the same time so many distinguished specialists on air law and space law. Our organization is, as you all know, a fully nonpolitical scientific organization, and this meeting is the platform for a free scientific interchange of ideas between lawyers. The fact that so many famous scholars have joined the meeting is the best guarantee for its success.

We have chosen to concentrate on two topics rather than to start a general discussion on space law. The two subject-matters we will deal with today are:

1. *International Control of Outer Space.*

2. *Damage to Third Parties on the Surface Caused by Space Vehicles.*

You can not really describe those two topics as problems of tomorrow. They are not even new today. In the light of space activities already performed they should by now have been solved by the lawyers. Nevertheless, they are still open to discussion and it seems to be an urgent task to reach a fair solution. It is the sincere wish of the Committee responsible for the arrangements of this meeting that the following discussions really might contribute to the clarification of some important points on space law.

With these words, Ladies and Gentlemen, I declare the Third Colloquium on the Law of Outer Space opened.

The Colloquium elected with acclamation Professor Grönfors as its Chairman.

Chairman: I thank you very much for the honour. I will try to fulfil my duties to the best of my abilities.

First I give the floor to *Dr. Jenks* for his *Introductory lecture* concerning International Control of Outer Space.

The International Control of Outer Space

Introductory Lecture
C. Wilfred Jenks

My task this morning is to pose questions rather than to answer them. I shall pose them from the angle of one who cannot claim any technical knowledge of astronautics, but who approaches the problem of the international control of outer space as a new and increasingly vital element in the general problem of effective world organization to secure the freedom and welfare of the human race by sustained economic and social progress on the basis of universal law. This is our assignment, not for this morning alone, but for a generation or more to come.

I do not intend on this occasion to enlarge on the fundamental astronomical and other scientific facts, the underlying political issues, or the more hypothetical legal problems. My views on these matters [1], and the views upon them which many of you have expressed from time to time, are on record elsewhere. We should, I submit, endeavour to address ourselves this morning to a more immediate issue, namely, how can we break the deadlock which at present bars the way to effective international control of space?

There is now an abundance of literature, much of it somewhat speculative, concerning space law [2], and in Jessup and Taubenfeld's *Controls for Outer Space* [3] we have a thoroughly scholarly examination of the whole problem by an international lawyer of the first rank.

An increasing number of responsible bodies, scientific, legal, and official, national and international, have considered or are considering the matter.

The initiative has come, as is meet, from astronauts, scientists and technologists.

The International Astronautical Federation has pioneered the whole venture and, in addition to sponsoring the Space Law Colloquia of which this is the third, established yesterday an International Institute of Space Law. We would, I am confident, all wish to give expression this morning to our keen satisfaction at this important departure and our ardent hope that the Institute will make a major contribution to future developments. It is designed to be an objective and scientific body; in order that it may fulfil its purpose it is essential that it should be fully representative and that all its work should be most carefully prepared in the closest consultation with leading scientists and technologists; we all trust that it will soon become fully representative in composition and character.

The International Council of Scientific Unions has throughout played a leading part, has to its credit the positive practical achievement of the substantial measure of international co-operation in space exploration achieved during the International Geophysical Year, and continues, through its Committee on Space Research (COSPAR) to stimulate and co-ordinate scientific co-operation in the whole field [4].

International lawyers have not been slow to explore the challenge which the new scientific and technological developments present for the law, but their collective thinking on the subject is necessarily in a tentative stage of development.

3

The International Law Association, which has for many years had an Air Law Committee, now has a special Sub-Committee on Air Sovereignty and the Legal Status of Outer Space; last week at Hamburg the Forty-Ninth Conference of the Association affirmed the principle that outer space may not be subject to the sovereignty or other exclusive rights of any State [5].

The Institute of International Law, which is properly a cautious body, now has a Commission on Space Law of which I have the honour to be chairman and reporter [6]. This Commission has not yet reached any conclusions.

While the scientists are playing, and the lawyers are preparing themselves to play, their respective parts, the dominating fact in the present situation is the increasing preoccupation with the problem of statesmen, governments, and official international organizations. Of this there have been many indications.

During the recent disarmament negotiations both sides have accepted, subject to varying conditions which have not yet been reconciled, the general principle that outer space should be used exclusively for peaceful purposes [7], but no clear definition of what is meant by peaceful purposes has yet been agreed.

The General Assembly of the United Nations established in 1958 an *Ad Hoc* Committee on the Peaceful Uses of Outer Space. While the Committee was handicapped by the refusal of the USSR and certain other States to take part on the basis of the composition of the Committee decided by the General Assembly, it adopted a most useful report [8] which, in addition to giving a convenient survey of the potential scope of space activities, the support facilities necessary for the effective conduct of space activities, and the existing arrangements for international co-operation, contains an abundance of practical suggestions concerning matters which should be further pursued. The report distinguishes between matters in respect of which international agreements providing for an "open and orderly conduct of space activities" can "form the basis of an international routine" without continuing co-operative action and matters in respect of which "there is need for active co-operative endeavours in which groups of nations assist each other in carrying out various phases of space activities". It envisages international agreements dealing with such matters as the use of radio frequencies, the registration of orbital elements, the termination of radiotransmission at the end of the satellite's useful life, the removal of spent satellites, the re-entry and recovery of space vehicles, the return of recovered equipment, identification of origin, and measures to minimize the adverse effects of possible biological, radiological and chemical contamination. Among the measures of international co-operation in joint projects contemplated by the report are simultaneous sounding rocket launchings, the international use of launching ranges, co-operation in the instrumentation of satellites and deep space probes, co-operation in tracking, telemetering and data processing, co-operation in the exchange and interpretation of data, and international arrangements to permit of a maximum use of communications and meteorological satellites. The Committee grouped as legal problems susceptible of priority treatment the questions of freedom of outer space for exploration and use, liability for injury or damage caused by space vehicles, the allocation of radio frequencies, the avoidance of interference between space vehicles and aircraft, the identification and registration of space vehicles and co-ordination of launchings and arrangements governing the re-entry and landing of space vehicles; it classed as less urgent the question of determining where outer space begins, the provision of safeguards against contamination of outer space or from outer space, questions relating to the exploration of celestial bodies, and the avoidance of interference among space vehicles. While these proposals need wider support to make them practicable, they represent an invaluable point of departure.

In 1959 the General Assembly established a new Committee on the Peaceful Uses of Outer Space, constituted on a broader basis, with a mandate to review the area of international co-operation in the exploration and exploitation of outer space for peaceful purposes, to study practical and feasible means for giving effect to programmes on the peaceful uses of outer space which could appropriately be undertaken under United Nations auspices, and to study the nature of the legal problems which may arise from the exploitation of outer space [9]. It has not yet been possible, however, to hold a meeting of this Committee with the participation of all its members.

The General Assembly has also approved in principle the convocation under the auspices of the United Nations, probably in 1961, of an international scientific conference for the exchange of experience in the peaceful uses of outer space, analogous in general character to the scientific conferences which have played so important a part in promoting the peaceful uses of atomic energy [10]. The detailed arrangements for the conference are still, however, in suspense.

The two specialized agencies of the United Nations most directly concerned with immediate developments, namely the International Telecommunication Union and the World Meteorological Organization, have already taken significant action. The International Telecommunication Union has allocated frequency bands for remote control of space vehicles; it has under study future needs for these purposes, the identification of radio-emissions from space vehicles, codes for the transmission of information from space vehicles, methods of providing for the cessation of radio emissions from space vehicles, and the use of space vehicle relays to extend terrestrial telecommunication facilities; and it has made arrangements for an Extraordinary Administrative Radio Conference to be held if timely in 1963 to consider certain of these matters further [11]. The World Meteorological Organization is playing an active part in advising on the use of artificial satellites for meteorological purposes and has envisaged that its future responsibilities might include participation in the planning of space vehicle tracking stations for meteorological purposes and in the design of the necessary computational practices and technique for the reduction of the data to amenable forms for practical use [12].

The International Civil Aviation Organization is vitally concerned with ensuring that the rules of the road to and from space do not conflict with the rules of the road in the air; this may imply that they should be framed and administered by the same international body; it certainly requires co-ordination on an international scale between space launching and aviation authorities comparable to that which it is believed already exists on a national basis in certain States where the matter has become of practical importance.

The Administrative Committee on Co-ordination, which consists of the Secretary-General of the United Nations and the executive heads of all the specialized agencies, has, at the request of the Economic and Social Council, made a preliminary examination of the steps desirable to promote co-ordination among all of the organizations of the United Nations family which are or may be concerned in the peaceful uses of outer space, and has the whole matter under continuing review [13].

This widespread preoccupation with the matter is encouraging, but it is disconcerting that few positive steps have yet been taken towards the establishment of effective arrangements for the international control of outer space. This is partly because one basic condition of the establishment of an effective international régime has not yet been fulfilled, partly because four major initial dilemmas which must be confronted before substantial progress can be made have not yet been resolved. Let us this morning attempt to visualize clearly the roadblocks which lie

5

before us, and then consider along what lines a sufficiently far-reaching practical programme can be evolved with some reasonable prospect of success.

The basic condition of the establishment of a satisfactory international régime which is still lacking is a sufficiently widespread distribution of space capabilities. There are clear indications that space capabilities are now likely to be more widely distributed in the near future; that a larger number of States will have at their disposal launching facilities provided by their own efforts, alone or in combination, or made available by the co-operation of States having such facilities; and that a still larger number of States will co-operate in the provision of instrumentation and in undertaking and appraising the results of scientific experiments. This wider distribution of space capabilities may be expected to have a twofold impact; while increasing the urgency of devising adequate international arrangements it should increasingly create the distribution and balance of effective participation in acti- vities in space necessary to sustain a satisfactory international régime; in so far as the wider distribution of space capabilities is achieved by combined action by a number of States it will also tend to create directly possible instrumentalities and other significant elements of a broader international régime.

Once this basic condition has been fulfilled we will still be confronted with four major initial dilemmas. How can we, despite the potential military implications and applications of most if not all activities in space, evolve, in the absence of an agreed international control and inspection of armaments which still eludes us, arrangements for peaceful co-operation in space which will reduce the danger that space will be so militarized before general arrangements for the international control of armaments have been agreed as to make it still more difficult to reach agreement upon such arrangements? How can we, without fettering the progress of science and technology or unnecessarily importing into the scientific field poli- tical considerations and rivalries or the rigidity of legal rules conceived in the study rather than evolved from practical experience of space activities, create a part- nership of policy, law, science and technology in bringing activities in space within the rule of law? How can we, without crystallizing policy prematurely on the basis of quite inadequate knowledge of future probabilities and possibilities, avoid the imminent danger that the pace of scientific and technological development will so outrun that of effective international control that events rather than policy will crystallize the pattern of the future in a manner which gives national rivalries a new dimension in space and makes the establishment of effective international control infinitely more difficult? How can we combine vigorous initiative in the international promotion and control of particular activities in space which permit at a relatively early stage of a high degree of international co-operation or control with the importance of ultimately co-ordinating particular activities within an appropriate framework of over-all international control?

All four of these dilemmas have one major feature in common. None of them can be resolved by preferring one of two clearly contrasted alternatives to the other. All of them call for an adjustment of the conflicting elements in the dilemma in such a manner as to enable us to make substantial progress without resolving ques- tions which are likely to remain insoluble either permanently or for some consider- able time to come.

Let us consider the type of adjustment which may be possible and necessary in respect of each of these dilemmas.

Our first dilemma is to reconcile the needs of peace with the dangers of war, the claims of scientific co-operation against the claims of military security. It represents an insoluble problem unless we can create a larger measure of international con- fidence — a measure of confidence to which the elimination of barriers to the full interchange of scientific information relating to and arising from activities in space, the pooling of certain activities in space, and guarantees of mutual security such as provision for the advance notification of launching and an inspection system can all make an important contribution. This dilemma is but one of many illustrations of the interdependence of all our problems. We cannot solve any of them unless we attempt to grapple simultaneously with all of them, but it is equally true that we cannot hope to solve any of them if we are unwilling to advance in any field until satisfied that the necessary correlative action has already been taken in other fields. We must therefore find lines of advance which represent a possible breakthrough from stalemate.

Our second dilemma is to determine the respective rôles of science, political action and law in international co-operation in space. International scientific co-operation of the type exemplified by the International Geophysical Year and the work of the International Council of Scientific Unions has initiated international co-operation in space with an encouraging measure of success. Why, then, cannot we as lawyers leave to the scientists the whole or major responsibility for the success of international co-operation in space? We may presume, and must pray, that the scientists know what they are about. Why must we display the lawyer's foible for meddling with things which he does not fully understand? This general attitude finds no countenance in the authoritative pronouncements of the Committee on Space Research (COSPAR) of the International Council of Scientific Unions which has recognized "the need for international regulation and control of certain aspects of satellite and space probe programmes", but it is sufficiently widespread among scientists and technologists to call for some comment. Certainly no lawyer should indulge in abstract speculation on the subject without first familiarizing himself with the scientific background and outlook [14]. But no scientist should cherish the illusion that space affords a new opportunity for the withering away of the State. The existing space programmes are government ventures; the launching facilities which have made them possible have been created and are operated by the military; we cannot hope to avoid confusion and frustration or to keep power politics out of space unless scientific co-operation and political understandings are translated into firm legal obligations. The only solution for the dilemma is a close partnership of policy, law, science and technology in determining the proper scope and content of such obligations.

Our third dilemma is to weigh the possible dangers of over-hasty action against the certain danger of excessive caution; we can resolve it only by distinguishing between matters in respect of which early action is imperative and matters which can and should be left for more mature consideration in the light of fuller knowledge and experience.

Similarly, the dilemma presented by the need to secure vigorous thrust in particular ventures in co-operation in space without prejudice to the long-term need for over-all co-ordination can be resolved only by an empirical approach in which both elements in the problem are kept in mind in the planning of such ventures.

One of the major unknown factors in the situation is the relative importance of activities in space as compared with, and the closeness of their relationship with, other scientific and technological developments which may call for some form of international co-operation, regulation or control. The dramatic character of activities in space and the novelty of the problems which they pose have rightly captured the public imagination, but nuclear transformations on earth, the wider use of electronics and automation, supersonic travel speeds, developments in the life sciences which may bring the creation of life itself and of new forms of life within the range of human action, fuller exploration of the depths and floor of the

7

ocean and the last remaining inaccessible areas of the earth, notably Antarctica and the highest Himalayas, deeper penetration of the earth's surface, and the piercing of the moho may produce changes in man's environment with both a more immediate and a more far-reaching impact, for good or for ill, than activities in space. We must therefore be careful not to lose sight of the total picture. While vision and boldness are essential, perspective and proportion are no less fundamental.

In approaching the problem we must at all times recall that activities in space are only one element in "the challenge presented by the fact that man's experiments with his natural environment have now reached a scale calling for a reasonable measure of international supervision·and control" [15]. Nor can we afford to forget that without humility we may rank in the story of the quest for and diffusion of knowledge with Icarus rather than with Prometheus.

We must now proceed from generalities to specifics. How can these general considerations of policy be translated into an effective programme of action?

Let us deal in turn with the problems presented by the fact of instruments in space and by the prospect of man in space.

Instruments in Space

What kind of international régime do we require for the effective control of instruments in space?

At the present stage of development earth satellites and space probes are primarily tools of scientific research. We have already learned much from them, and may reasonably hope to learn much more, in regard to such matters as the shape and detailed measurement of the earth, the density, composition and temperature of the upper atmosphere, irregularities in the atmosphere (including variations of density, winds, and the vagaries of the ionosphere), the zones of radiation which satellites have detected high above the earth, the magnetism of the earth, cosmic rays, the radiation of the sun, the traffic density of meteors, and the effects of weightlessness on animals. This information has been secured by four methods: by tracking the orbits of satellites and space probes, optically and by radio and radar; by radio-recovery of the readings of satellite and space probe borne instruments, frequently made more valuable by telemetering and electronic memory devices; by radio photography; and, most recently and most far-reaching in its potential implications, by the recovery of data capsules ejected from satellites. Orbit tracking supplemented by the mathematical interpretation of the data obtained thereby has been particularly important as a source of geodesical information and has also thrown light on the basic properties of the atmosphere; for other purposes the other methods have been, and will increasingly be, more important. Orbit tracking can be undertaken by any State within reach of which a satellite or space probe passes which has the necessary organization and scientific equipment; information from radio transmissions and radio photography by satellites and space probes is available directly only to States which can receive and interpret it; data capsules afford information directly only to States which recover them and can interpret their contents.

Three series of further developments in the use of instruments in space may be anticipated in the relatively near future.

Improved instrumentation, which may include satellite-borne telescopes, spectroscopes and other instruments, will make possible a wider variety of new and more complex observations and experiments and the recovery of fuller and more reliable data.

The range of space exploration by instruments may be expected to be extended by the placing in orbit of artificial satellites of the moon, the sun, the nearer planets and possibly certain comets, and the maintenance of contact with them for substantial and increasing periods.

The use of instruments in space will pass from the experimental to an operational phase with the placing in sufficiently stable orbits of a number of communications, meteorological, navigation and astronomical observation satellites adequate to permit of dependable telecommunication relays, meteorological services, navigational aids, and space observatories.

These constructive possibilities may, however, be overshadowed at any time by the use of space for military purposes. We are told, for instance, that it is technically possible to place nuclear weapons in orbit and release them by a radio signal.

In this general situation what rôle has the law to play and what measures of international control do we require?

Unless space is exclusively dedicated to peaceful and scientific purposes, and a satisfactory inspection system designed to make the exclusive dedication of space to peaceful and scientific purposes effective is introduced, all else will be precarious. This is primarily a problem of the control of armaments rather than a problem of the control of space.

Pending a satisfactory outcome of negotiations for disarmament, what other steps can usefully be taken?

In the first place, legal regulation is a necessary element in the measure of common discipline without which the scientific exploration of space cannot proceed fruitfully. This is most obviously so in respect of space telecommunication, a field in which the International Telecommunication Union has already taken, and must continue to hold, the initiative. It is equally so in respect of measures to minimize the adverse effects of possible biological, radiological, and chemical contamination of and from space, a matter in respect of which the initiative lies for the moment with CETEX, the Committee on Contamination in Extra-Terrestrial Exploration of the Committee on Space Research of the International Council of Scientific Unions, but which may well call for legal regulation to buttress understandings among scientists. Unless clear rules on these matters exist from the outset and are strictly applied, space research will not yield the fruits which we are entitled to expect from it. Over a longer period it may be equally important for the same purpose to have clear rules requiring space vehicles to be fitted with instruments permitting their recovery or destruction at the end of their useful lives in order to avoid space being cluttered with derelicts the continued observation of which places an impossible burden on the ground tracking networks.

These prospective measures represent a minimum of international co-operation without which nobody will be able to begin to reap efficiently the fruits of the scientific exploration of space; a much larger measure of co-operation will be necessary before anybody can reap the full harvest, the garnering of which involves simultaneous observation in many parts of the world; a still larger measure of co-operation will be necessary to make that harvest available to everybody in such a manner as to eliminate national rivalry from space.

Simultaneous observation in many parts of the world may be partly the spontaneous result of scientific or military curiosity and partly the outcome of understandings among scientists and observational stations, but its successful organization on a long-term basis calls for an agreed code of rules providing as a minimum for the exchange of full tracking information and of methods of data decoding and ideally for a much more complete exchange of information, including a complete tabulation of reduced, calibrated and corrected data and facilities for verifying matters calling for further enquiry by access to telemeter records.

The further stage of making the full harvest of new knowledge available to everybody has a major bearing on the demilitarization of space. Mutual protection against surprise attack is the key to making effective the exclusive dedication of space to peaceful purposes; the chief danger of activities in space unleashing war on earth may well lie in some inoffensive space vehicle being mistaken in a radar screen at a moment of heigthened international tension for an intercontinental ballistic missile which has been launched for a military purpose. Advance notification of launching sites and firing schedules, the filing of flight plans and of descriptions of weight, load and size, and the use of agreed radio codes for the reception of data from space can all play a significant part in eliminating the military element. The international arrangements desirable for this purpose may well include a special organization analogous to that which is projected for the detection of any violation of the proposed nuclear test ban.

The use in artificial satellites and other space instrumentalities of observational equipment involves a major danger of international incident in the absence of clear understandings expressed in the form of clear-cut obligations. The solution of the problems to which it gives rise may perhaps be found in a requirement that information obtained by the use of such equipment shall not be reserved for the exclusive use, whether for military or for other purposes, of the State having collected it, but shall be made available to the whole world. Example may be more effective than precept as a means of inaugurating such a practice, but the practice is not likely to be generally observed despite national pressures to maintain secrecy unless it is fortified by accepted international obligations covering the matter.

The elimination of the military element has a special significance for States which are preoccupied by the possibility that the passage of space vehicles through their atmosphere may be regarded by more powerful States as a breach of neutral duty. While it would appear to be clear that no State can reasonably be held responsible for failure to prevent something which only the most advanced and expensive technology, as yet within the resources of only two powers, can either achieve or prevent, an international agreement which would eliminate the problem is eminently desirable.

The launching, flight and re-entry of space vehicles involve a number of dangers to third parties — dangers from misfires, dangers to aircraft, and dangers from loss of control on re-entry. The question of liability for injury or damage caused by space vehicles is therefore one of high priority. It includes the principle and extent of liability (a problem the solution of which may be found in a combination of absolute liability for injury or damage on the surface or in the air and liability based on fault for injury or damage in space), the question whether liability should be unlimited in amount and if not what limitation of liability should apply, the provision of insurance against liability, and possibly, in certain cases, the question whether liability is joint or several. The possibility of a serious international incident arising from injury or damage caused by a space vehicle cannot be ignored and the question is therefore of urgent political as well as of legal importance. An early international agreement settling the law on the subject, establishing a procedure for the recovery of damages which gives injured persons an effective means of redress, and providing for the submission to the International Court of Justice of disputes between States concerning the liability of States for injury or damage caused by space vehicles is therefore desirable. Mr. Pépin will be dealing with this matter this afternoon and I shall therefore not discuss it further.

Operational activities in space for telecommunication, meteorological, navigation and astronomical purposes will call as a minimum for codes of regulations governing their operations, but they may also present an opportunity to organize such activities as world public services rather than as national or commercial under-takings. The International Telecommunication Union, the World Meteorological Organization and the Intergovernmental Maritime Consultative Organization are not at present operational bodies, but the possibility of entrusting them with certain operational responsibilities in connection with the operation of such world public services should not be lightly set aside. If provision for the discharge of such respon-sibilities cannot be made within the normal structure of the existing specialized agencies of the United Nations, special bodies, similar to the European Council for Nuclear Research, should be envisaged for the purpose, preferably within their general framework; alternatively, a body with a more comprehensive responsibility, analogous in general conception to the International Atomic Energy Agency but with more specifically operational responsibilities, might be created for the purpose within the general United Nations framework and some provision made for the co-operation with it of the existing specialized agencies. Large commercial interests have been involved for some considerable time in the design and manufacture of space equipment of all types and have a heavy investment therein. It is now re-sponsibly reported that, in the United States at least [16], the determination of the future relationship between government and industry in the utilization of space has become a major preoccupation of policy for the immediate future, involving such questions as the extent to which operational functions in space should be dis-charged by government or private industry, the extent to which government launching facilities, satellites, or research or development facilities should be avail-able to private industry, and the extent to which particular private concerns should be allowed to secure a favoured or monopolistic position in respect of particular space services. These are large, highly controversial and urgent questions directly affecting far-flung and powerful economic interests and having far-reaching poli-tical and cultural implications. The opportunity to organize operational activities in space as world public services will be a fleeting one but every effort should be made to grasp it and to exploit it to the full. Success in this respect would be a major contribution to the establishment of an effective international control of space.

Man in Space

While adequate international arrangements for the control of instruments in space represent the most urgent problem and would constitute a major advance, they will be liable to be outdated overnight by the appearance of man in space.

It is now widely and responsibly believed that we are on the threshold of man venturing into space [17]. How many further political and legal problems man's venture into space will bring with it cannot be foreseen with any accuracy until we know much more of the nature and extent of his probable activities in space. Scientific expeditions, military activities, and exploitation of the resources of space in a manner still undefined would create entirely different situations and problems; brief ventures into space, continuous residence in space for substantial periods, and permanent residence in space with reproduction of the species there would likewise involve wholly different problems. We can perhaps envisage three successive phases, which for purposes of convenience I propose to describe as the Monroe Doctrine for the Moon phase, the Antarctic Analogy phase, and the United Nations control phase.

The first phase will presumably be one of hazardous and intermittent exploration with at most a handful of people in space at any one time. I have suggested in *The Common Law of Mankind* [18] that for this phase we need a Monroe Doctrine for the Moon concept. The expression is of course a metaphor and I recognize that the term may be unacceptable in view of its historical associations with one of the space powers, but the concept of hands off which it conveniently expresses is a fundamental one and its validity does not depend on the acceptability of the term. In this first phase we cannot reasonably hope to have evolved either adequate institutional arrangements or sufficiently fully accepted legal principles to meet longer-term needs, nor will it be practically necessary that we should have done so, but it will be vital to have established the position that such arrangements and principles will be evolved by common agreement on the basis of growing experience as they become necessary and that meanwhile no unilateral claim to extraterrestrial sovereignty or jurisdiction or to exclusive access to or use of any extraterrestrial place or resource will be recognized. An immediate declaration of policy to that effect, preferably by the General Assembly of the United Nations, is eminently desirable. The Monroe Doctrine for the Moon concept represents a holding operation designed to ensure that longer-range plans are not prejudiced by unilateral action before they have had time to mature.

Such longer-range plans also need immediate consideration, but as it is desirable that they should be embodied in treaty obligations they will presumably take longer to mature. In formulating them we may usefully be guided by the Antarctic analogy. The usefulness of the Antarctic analogy has been fully discussed in Jessup and Taubenfeld, *Controls for Outer Space and the Antarctic Analogy,* and I therefore do not propose to recapitulate their analysis and argument, but since their book was published the Antarctic Treaty of 1 December 1959 [19] has been signed on behalf of all twelve of the countries which participated in the Antarctic programme of the International Geophysical Year. The Antarctic Treaty recites that it is in the interest of all mankind that Antarctica shall continue forever to be used exclusively for peaceful purposes and shall not become the scene or object of international discord, and that the establishment of a firm foundation for the continuation and development of international co-operation in scientific investigation in Antarctica on the basis of freedom of scientific investigation as applied during the International Geophysical Year accords with the interests of science and the progress of all mankind. The Treaty provides that Antarctica shall be used for peaceful purposes only and that any measures of a military nature, such as the establishment of military bases and fortifications, the carrying out of military manoeuvres, and the testing of any type of weapons shall be prohibited. Freedom of scientific investigation in Antarctica and co-operation toward that end, as applied during the International Geophysical Year, are to continue. The Contracting Parties agree that, to the greatest extent feasible and practicable, information regarding plans for scientific programmes in Antarctica shall be exchanged to permit maximum economy and efficiency of operation, scientific personnel shall be exchanged in Antarctica between expeditions and stations, and scientific observations and results from Antarctica shall be exchanged and made freely available. In implementing these principles every encouragement is to be given to the establishment of co-operative working relations with those specialized agencies of the United Nations and other international organizations having a scientific or technical interest in Antarctica. Territorial claims are frozen without prejudice for or against for the duration of the Treaty. Any nuclear explosions in Antarctica and the disposal there of radioactive waste are prohibited. Each Party may designate observers who are to have complete freedom of access at any time to any or all areas of Antarctica; all stations, installations, equipment, ships and aircraft are to be open at all times to inspection by such observers and aerial observation may be carried out at any time; such observers and their staffs are to be subject only to the jurisdiction of the Party of which they are nationals while they are in Antarctica for the purpose of exercising their functions. There is to be advance notification of all expeditions to and within Antarctica. There is provision for periodical meetings of representatives of the Contracting Parties who may recommend to their governments measures in furtherance of the principles and objectives of the treaty. The negotiation at an early date of a similar treaty concerning space, based on the same general principles but containing any necessary adaptations, appears to be desirable. It may be desirable to formulate at an early stage in the proposed treaty or otherwise a legal duty to give mutual aid against the common dangers of the unknown and proceed to the assistance of persons in distress. The principle is fundamental in the ethics of mountaineering and polar exploration and has already received legal expression in the Safety of Life at Sea Convention and the Search and Rescue Annex to the International Civil Aviation Convention. A Space Treaty analogous to the Antarctic Treaty would represent a bold and constructive advance. If such a Treaty is possible for Antarctica, why is it not possible for Space?

As activities in space develop, arrangements analogous to those provided for in the Antarctic Treaty may prove to be inadequate and more comprehensive arrangements for international control of space may be necessary. We must therefore envisage a third phase, that of United Nations control. The type of institutional arrangements appropriate for such control must be determined in the light of how the United Nations has developed when that phase is reached, but there are certain general principles which it is not premature to foreshadow now.

We may start from the general point of departure defined by Jessup and Taubenfeld:

> "It is clear on a moment's reflection that now, and probably for some time to come, many of the problems connected with outer space are more closely connected with earth and with man on earth. Earth is still the launching site and the base for sending and receiving communications. The needs or desires which inspire the exploration of outer space are earth-born and earth-centered. The use or misuse of outer space of which we speak is man's use or misuse. In speaking of "controls for outer space" we are thinking of man-made and man-applied controls; of controls not *of* space, but of man-made objects and also of men *in* space, and ultimately of men on planets in space" [20].

When man ventures into space he takes with him much of his earthly heritage, including the established rules of international law in so far as they are applicable. The Charter of the United Nations is not earthbound. The General Treaty for the Renunciation of War and the Statute of the International Court of Justice, "international custom as evidence of a general practice accepted as law" and "the general principles of law recognized by civilized nations" are all applicable to human relations in space. Within this framework the special political and legal arrangements necessary for the effective international control of space when the intensity and complexity of space activities has passed beyond the stage at which arrangements analogous to the Antarctic Treaty are adequate can and must be evolved.

Whither Now?

I began by saying that my task this morning was to pose questions rather than answer them and I must conclude upon the same note. Let me recapitulate briefly six questions which call for an answer.

1. *Why cannot we negotiate forthwith a network of international agreements dealing with such matters as space telecommunications, the avoidance of contamination of and from space, mutual aid in simultaneous observation, mutual protection against surprise, the use in space of observational equipment, the freeing of*

space information from security restrictions, and injury or damage caused by space vehicles, to be followed by further agreements as required?

2. Why cannot we, while there is yet time, organize space telecommunication relays, space meteorological services, navigation satellite systems and space observatories as world public services maintained under United Nations auspices in the interest of all mankind?

3. Why cannot we, before scientific adventure places in jeopardy the existence of human life itself, provide machinery for the notification of, and exchange of information concerning, experiments, tests and development schemes, in space or elsewhere, which are liable to affect the natural environment of other States and possibly of the whole human race?

4. Why cannot we immediately, before man is in space, promulgate the principle, perhaps in a Declaration by the General Assembly of the United Nations, that no unilateral claim to extraterrestrial sovereignty or jurisdiction or to exclusive access to or use of any extraterrestrial place or resource will be recognized?

5. Why cannot we negotiate at an early date a Space Treaty, analogous in general character to the Antarctic Treaty, providing that space shall "continue forever to be used exclusively for peaceful purposes and shall not become the scene or object of international discord" and embodying the arrangements necessary to this end?

6. Why cannot we lay the foundations now for a comprehensive plan of United Nations control of all earth-based and earth-directed activities in space?

Upon the answer which we give to these questions may depend not only the future of man in space but equally the future of man on earth.

References

[1] The Common Law of Mankind, pp. 382—407, (London, Stevens and Sons Ltd.; New York, Frederick A. Praeger Inc., 1958), which reproduces with some additional suggestions my paper on "International Law and Activities in Space", originally published in 5 International and Comparative Law Quarterly, 1956, pp. 99—114.

[2] Most of the early papers on the subject, from the pioneer work of *John Cobb Cooper* and the *Prince of Hanover* to late 1958, are available in the invaluable Symposium on Space Law, published 1958 (United States Senate, Eighty-Fifth Congress, Second Session, Special Committee on Space and Astronautics, Space Law — A Symposium, prepared at the request of Hon. Lyndon B. Johnson, Chairman, December 31, 1958); a number of the more recent papers were read before the Space Law Colloquia held by the International Astronautical Federation in 1958 (*Andrew G. Haley* and Dr. *Welf Heinrich, Prince of Hanover,* First Colloquium on the Law of Outer Space, The Hague, 1958, Proceedings, Wien, Springer-Verlag, 1959); and 1959 (Second Colloquium on the Law of Outer Space, London, 1959, Proceedings, Wien, Springer-Verlag, 1960). Elaborate bibliographies have been published by Mr. John C. Hogan, Dr. Eugène Pépin, and others. See, for instance, *John C. Hogan,* Space Law Bibliography, 23 Journal of Air Law and Commerce, 1956, pp. 317—325, and A Guide to the Study of Space Law, 1958; *Eugène Pépin, Bibliographie des Travaux Publiés sur les Problèmes Juridiques de l'Espace et Questions Annexes,* 1910—15 septembre 1959 (Extrait de la Revue Française de Droit Aérien, No. 4 de 1959). Lectures have been given on the subject at the Hague Academy of International Law. See *Rolando Quadri, Droit international Cosmique,* in Recueil des Cours de l'Académie de Droit

International, Vol. 98, 1959 (III), pp. 509—597. The Rockefeller Foundation has sponsored a substantial programme of research on the subject.

[3] *Philip C. Jessup* and *Howard J. Taubenfeld,* Controls for Outer Space and the Antarctic Analogy (New York, 1959).

[4] See I.C.S.U., A Brief Outline (1958).

[5] International Law Association, Report of the Forty-Eighth Conference (New York, 1958), pp. 246—271 and 320—330; Report of the Forty-Ninth Conference (Hamburg, 1960, still in the press).

[6] *Annuaire de l'Institut de droit international, 48,* II (1955), pp. 357 and 499; the report of the Commission will appear in a later volume of the *Annuaire.*

[7] United Nations Document D.C./179 of 15 August 1960.

[8] United Nations Document A/4141 of 14 July 1959.

[9] General Assembly Resolution No. 1472 (XIV) A of 12 December 1959.

[10] General Assembly Resolution No. 1472 (XIV) B of 12 December 1959.

[11] Twenty-Fourth Report of the Administrative Committee on Co-ordination to the Economic and Social Council, United Nations Document E/3368 of 10 May 1960, Annex I, paragraphs 7 to 17.

[12] *ibid.,* Annex I, paragraphs 18 to 25.

[13] Twenty-Fourth Report of the Administrative Committee on Co-ordination to the Economic and Social Council, United Nations Document E/3368 of 10 May 1960, Part V, paragraphs 18 to 21.

[14] *e.g.,* with the history of rocketry as recorded in *Andrew Haley,* Rocketry and Space Exploration (1958); with such scientific works as *Van Allen,* Scientific Uses of Earth Satellites (2nd ed. 1958), *Massey* and *Boyd,* The Upper Atmosphere (1958), *King-Hele,* Satellites and Scientific Research (1960); and *Ari Shternfeld,* Soviet Space Science (1959); see also Soviet Writings on Earth Satellites and Space Travel (London, Macgibbon and Kee, 1959); *M. Vassiliev,* Sputnik into Space (1958); USSR Academy of Sciences, translated by *J. B. Sykes,* The Other Side of the Moon (Pergamon Press, 1960); and *Konstantin Tsiolkovsky,* Beyond the Planet Earth (Pergamon Press, 1960), which, though a translation of a fantasy by the pioneer of Russian rocketry published in 1920, is illuminating; with the Rand Corporation's comprehensive Space Handbook — *Robert W. Buchheim* and the Staff of the Rand Corporation, Space Handbook: Astronautics and its Applications (1959) — originally published as United States Congress, 85th Congress, 2nd Session, Space Handbook: Astronautics and its Applications, Staff Report of the Select Committee on Astronautics and Space Exploration (1959); and above all with the I.G.Y. Manual on Rockets and Satellites (Annals of the International Geophysical Year, Vol. VI, I.G.Y. Manual on Rockets and Satellites, edited by *L. V. Berkner,* 1958), supplemented for the more recent developments by the periodical Planetary and Space Science, issued by the Pergamon Press.

[15] I have discussed the matter more fully in *C. Wilfred Jenks,* The Laws of Nature and International Law, in Netherlands International Law Review, 1959, Liber Amicorum J. P. A. François, pp. 160—172.

[16] New York Times, International Edition, 26 July 1960, p. 1, col. 5.

[17] For a recent appraisal of the possibilities, see *Kenneth F. Gantz,* Man in Space — Principles and Practice of Space Flight as Developed by the United States Air Force (1959) with a Preface by General Thomas D. White, Chief of Staff, United States Air Force; see also *Hermann Oberth,* Man Into Space (1957).

[18] *C. Wilfred Jenks,* The Common Law of Mankind, p. 405 (1958).

[19] For the text see American Journal of International Law, Vol. 54, No. 2, April 1960, pp. 476—483, or United Kingdom Parliamentary Paper Cmnd. 913 of 1959.

[20] *Philip C. Jessup* and *Howard J. Taubenfeld,* Controls for Outer Space and the Antarctic Analogy, p. 4 (1959).

Discussion

Before the discussion the papers pp. 21—127 were presented.

Rinck

May I invite you to consider space law from a practical point of view? Let us leave aside for a moment the law as we want it to be and let us discuss instead the law as it stands to-day. We must be ready to answer one practical question any time. It is this: "Shall a state tolerate being overflown by satellites?" Putting the same problem somewhat differently, the point is, how far does national sovereignty extend into space?

In order to answer this question methodically we must first ask ourselves: what is the substance and object of sovereignty? Sovereignty is established by common consent with the object to protect a state or a nation. If a person owns or if a state rules over some land they want a little space to breathe and they want more space to be protected against interference from above. So it is an inherent element in sovereignty that it extends to some degree into space. I submit, Mr. Chairman, that sovereignty reaches as far as satellites can fly. Therefore no state is obliged to tolerate a satellite above its territory.

Teleologically conceived sovereignty means that no interference from outside is allowed. A satellite, any satellite in fact, is a potential danger to the territory underneath. We have heard this morning that satellites may carry nuclear warheads and that these may be released any time. No state is obliged to allow such a passage. Any satellite, no matter how high its orbit, is a potential danger to the subjacent state and is therefore incompatible with sovereignty. — That is what sovereignty commands, it must therefore be extended as high as satellites can fly, *i.e.* as far as the earth's gravitation works. Every satellite uses the force of gravitation. Where gravitation ends, the satellite flies off, it is no longer a satellite but becomes a true space craft and enters celestial space.

I am well aware that this borderline for sovereignty coinciding with gravitation is very far flung. It may extent up to a million miles as some scientists say. Leave the demarcation in detail to the specialists. International law says that as far as gravitation works sovereignty must be recognized. In theory gravitation never ends, we know that, but in practice it does. When gravitation from other celestial bodies (not from our moon) grows stronger than this planet's force, then the borderline of sovereignty is reached.

Not all satellites are dangerous. They may serve peaceful purposes, scientific research or inspection. Unarmed craft and satellites equipped for surveying and inspection only must be allowed to pass over all states, provided their innocent character is proved. But I think we want an international convention for that.

Within this Federation and these Colloquia we nearly always agree on the law as it should be. The bigger problem, however, may be the actual law. Here my thesis is: No state need tolerate the passage of an unknown vehicle. It therefore may try to disturb, deviate or destroy any unauthorized satellite.

Smirnoff

Mr. President, Ladies and Gentlemen. Today we heard the excellent paper, an introduction by Mr. Jenks, and I am in fully accord with him, that the six points he pointed out to be discussed at once are the minimum that we can do. The most serious remark made in this session to the points of Mr. Jenks was put by the pioneer of space law Professor Cooper and seconded by Mr. Hyman. It is remarked that the preliminary question which must be solved before going into the detailed discussion about these six points is the right delimitation of outer space and air space. At this moment I should like to say some words.

Gentlemen, I am not minimizing the significance of the remarks of Professor Cooper and Mr. Hyman. The Yugoslav delegation, which was in Hamburg on the ILA Conference, put this proposition which was taken into the resolution of this ILA Conference. But I think that *de lege ferenda* this moment is not so important because with the development of space flight and with the development of earth to earth flight through the space, air law as such and the problems of air law will lose their significance like the Montgolfier balloon lost its significance with the air law of the 20th century. Thus, I should like not to stop our

16

work for the question of delimitation of air space and outer space, which is a very, very difficult problem. As I consider it, this problem is a temporary problem and perhaps I could advance a proposition made by the valuable Brasilian jurists in the Hague in 1958, which was a negative one: Outer space is every place where the aircraft could not fly, including the jetplanes. With this definition you have the solution of many problems, for instance of problems of responsibility and so on. I do not insist on this definition because we have many perhaps technically better definitions like the one of von Kármán, like the three zones theory of the distinguished Professor Cooper, like the very interesting theory of Mr. Hyman put today with the neutral zone, and so on.

If I insist upon the continuing our work on the problems which Mr. Jenks put forward it is because I think that now the rôle of organizations like ours, like the ILA and others, has become very important in the space law problems. Two events in the later time have put in advance a very great significance of this branch of the law. One is — in my opinion, I am only speaking for myself — the declaration of a very important personality in Paris on the 16th of May, 1960, when the argument of classical air law was answered with the arguments of flying of satellites. The second event was on the 1st of June in the British Parliament when a conservative member of Parliament, Sir Richard Glyn, put the following question: "Which measures were taken by the British Government to bring down the Russian space ship, which is infringing the rights of Britain on its air or space?" The response of the British Minister was negative; no measures were taken. These two facts prove to me that the time has arrived when really we cannot stop for a very big technical and technological question of delimitation of outer space and air space in our works in space law problems.

Safavi

I believe the definition of outer space will afford some difficulties in respect of legal considerations, because the outer space cannot be other than the space itself. It would perhaps be more adequate to give a definition of the space and, then, to divide it in different sections. Therefore, there will be inner space and above certain limits outer space.

In the universe, there exist planets separated by space which we use for the purpose of communication and transportation. In fact, if we want to give a legal definition of space, we must base our definition upon the earth on which we live and from which we consider the whole universe. The earth has its own space which is under its attraction and, consequently, under its possession. So, if we put something into this space, it will normally fall down and will be submitted to the attraction force of the earth.

Are there any natural arguments better than this specific attraction of the earth to prove that there is one space for our planet with its own limits, which should be called terrestrial space? This terrestrial space begins from the surface of the earth and will end at the level in which there is no more attraction of the earth. Beyond this upper level, there is also a space which belongs to the other planets, but I call it extra-terrestrial space. The extra-terrestrial space has no upper limit and, therefore, I do not think it is advisable to give an upper limit for outer space. The terrestrial space belongs to the earth, that is to say, it belongs to all the nations existing on the earth.

Apart from the above mentioned limits, it is necessary to make one subdivision. It means: the part of the space existing above each country until a fixed level, a part which should be called territorial space, must be considered as a part of the country and submitted to its sovereignty. Here, I want to call your attention to a recognition of the sovereignty of the states upon the overlying airspace. This recognition was made by the Chicago Convention of 1944, which says, in Article 1: "The Contracting States recognize that every state has complete and exclusive sovereignty over the airspace above its territory." If the Chicago Convention used exclusively the term "airspace", the reason is that, at the time, only aircraft existed which could not move in the area beyond the air.

Territorial space, the upper limit of which must be, in my opinion, the limit of the air, should be considered as a part of the underlying country and submitted to the sovereignty of the state of the country. The upper limit of this territorial space is, in fact, the vertical boundary of each country. The way in and out by any kind of vehicles in this area must be made in accordance with an authorization of the government concerned, but beyond this vertical boundary up to the limit of terrestrial space, the transportation and the use of the space should be submitted to an international convention. Which takes into consideration, not only the problems of the space, but also the security of the earth and the vehicles with its passengers moving in that space.

If I consider the airspace as a part of each country and submit it to the sovereignty of that country, the reason is that the earth revolves with its own air and each country has a fixed airspace, which belongs to it, and it should have a vertical frontier which protects it against any intrusion. As the space above the airspace does not revolve with the earth, it is therefore impossible to fix national limits in that part of the Space. Therefore, that part of space belongs to all nations and should be regulated by an international convention. The need of a common use of that space for all nations and the security of the earth and vehicles and passengers therein make imperative the mutual collaboration of different nations.

According to that Convention, the transportation and circulation of all vehicles must be free, but submitted to certain rules and regulations of the space.

Beyond the terrestrial space or in the extra-terrestrial space, the freedom of the transportation and circulation should be recognized, but the rules and regulations, which must be observed, are rather specially concerned with the security of the spacecraft and passengers carried therein and with the hygiene in other planets.

Fasan

Considering the questions No. 4 and 5, put before us by Mr. Jenks, we must pay attention to the fundamental physical differences, which do exist between the nature of the empty (or nearly empty) space itself and the celestial bodies, moving therein. Even these bodies are divided into different groups:
1. Such with a firm surface, as the moon and the planets, and those without such a surface, as the sun and the other fixed stars.
2. Uninhabited bodies and such, inhabited by other intelligences. Though we do not know other inhabited bodies than our earth, we must consider this possibility, too.

The freedom of space (that means freedom from any sovereignty) and the legal impossibility of claiming regions of empty space is one of the few legal principles, upon which all jurists seem to have agreed.

But the question of sovereignty over celestial bodies has not received such an unanimous answer. The theories, advanced about the legal character of the stars, planets and moons, especially about our own satellite, can be summarized as follows:

1) Declaration of the moon as a free, independent and autonomous zone.
2) Free utilisation and exploitation of the moon like the high seas.
3) International administration of the whole satellite, for instance by a space agency, as it was discussed by the United Nations *Ad Hoc* Committee on the Peaceful Uses of Outer Space.
4) To draw an analogy to Antarctica and perhaps find a solution similar to the Washington Treaty of December 1, 1959.
5) The possibility of obtaining sovereignty over the moon *in toto* by an individual state.
6) The permissibility of occupying parts of the satellite by different states of the earth.
7) That the moon belongs periodically, and so possibly only for a few given minutes, to that country, in the zenith of which it just happens to be.

I remember that Mr. Haley once made a warning that if there is one nation landing on the moon and then perhaps will claim a part of the moon or the whole satellite then it will be difficult and there will be an endless embarrassing dispute. Therefore the questions number 4 and number 5 put before us by Dr. Jenks, are very urgent and ought to be dealt with and solved as soon as possible.

Gunnarsen

Ladies and Gentlemen. The thing that brought me forward was two remarks from two of the distinguished speakers before me. They mentioned that space or terrestrial space should extend to the point, where gravitation ends. I discussed this question with some of the persons now present at the Royal Institute of Technology here in Stockholm and they affirmed what was my impression from the schooling I had as a boy, that gravitation will never end and therefore this is an impossibility as fundament for a legal theory.

This is called a Colloquium of Space Law. It has developed into a colloquium of space politics. I do think too little has been spoken about the space law in existence. I am personally convinced that some rules exist already and I shall summarize those rules. There was a statement made from the Russian side shortly after the launching of the first artificial satellite saying that: "This satellite would not offend any territory of other states. Disputing the legality of the appearance of the satellite over the territory of one or another state would be just as absurd and ridiculous as disputing the appearance over a given territory of the moon, the sun or any other heavenly body." Now this is the official point of view from the Russian side. The Americans have stated through the Deputy United States Secretary of Defence: "I can only express the Defence Departments view of it, but if the Russians did place in orbit a satellite that had reconnaissance possibilities, we would consider that it was inoffensive in the sense that it was in outer space, where it could do us no harm and we would not object to it."

Now, those are two parallel official announcements from the only existing space powers legalizing or at least in practice telling us that satellites can move legally; in other words that they are legal where they move while in orbit. But this does not say anything about what happens below the area where the satellites can move as such. Now at the Royal Institute of Technology I have also heard, that the satellites cannot come any closer to the earth than roughly 90—120 miles. The answer to where national sovereignty ends must, according to my opinion, lay there in between.

Concluding Remarks of Session 1

C. W. Jenks

Mr. Chairman, Ladies and Gentlemen. Let me first thank everyone who has referred to what I had to say this morning for the much too generous language which they have used in describing what I have attempted to say. There is, I believe, one fundamental principle of all adventure in space. It is that one should never stay in space beyond the margin of safety allowed by one's existing resources. We are, I am afraid, at this juncture somewhat in danger of doing that. It is high time for Mr. Pépin to bring us back to earth.

But, before you do that, I should perhaps say one word about a suggestion made this morning by Mr. Andrew Haley, that an attempt should be made to formalize the conclusions to be drawn from this discussion in a resolution. I must confess that I had some little doubt this morning as to whether that might not be premature. And that doubt has been more than confirmed by the course of the discussion as the day has proceeded. I wonder whether it might not be wiser to follow an alternative course of action. A whole series of working groups has now been established by the Institute of Space Law to examine these matters more thoroughly. Would it not perhaps be wiser to have referred to these working groups the discussion which has taken place today, in the hope that they may be able to hammer out carefully and deliberately really worthwhile conclusions on the various points which have been raised?

There is, I think, very little for me to say by way of winding up this discussion.

I would like to refer, very, very briefly, to some few of the points which have been raised.

The question of demarcation is of course important. Of course it would be desirable to settle the matter. But can you settle it? We have had today two sharply contrasted suggestions. One: a limit of eighty to a hundred miles. The other: a limit of one million miles. And there are any number of possible intermediate positions. Surely we cannot afford to postpone dealing with problems of immediate practical urgency until we have settled what may prove to be a theoretical question. When for the purpose of settling problems of immediate practical urgency we have to get some tentative answer to that question for some particular purpose, then by all means let us do so. But it may be that we shall find that we have to give different answers for different purposes. And we may be wise, when framing an answer for a particular purpose, to frame it in such terms that it does not embarrass us in some unexpected context when we come to deal with an entirely different matter.

The right of self-defence has also figured in the discussion. And of course the right of self-defence is fundamental. But so are the limitations to the right of self-defence. As I understand the common law, as I understand the generally-accepted principles of international law, as I understand some of the statements which Professor Cooper quoted this morning, there are two main limitations to the right of self-defence. The first is that the measures taken must be reasonable in relation to the imminence and magnitude of the danger. And the second is that he who takes measures of self-defence is not the final judge of whether or not those measures are justified. If we depart from those principles the right of self-defense lets in everything, and there ceases to be any means of applying or enforcing the principle that space should be used only for peaceful purposes.

That leaves the other question raised during the discussion as to what is meant by peaceful use. It has been suggested that by peaceful use we mean any non-aggressive use. Again, surely it all depends on the essence of the matter in the case of the particular activity or transaction. They have in Canada a constitutional doctrine the essence of which is that in determining whether any particular subject of legislative jurisdiction falls within the competence of the Dominion or the provinces you have to penetrate to the substance of the matter and find out what the real purpose of the particular action is. Surely that is the only sort of basis on which you can approach this question of the definition of peaceful use. Because if you attempt to approach it on any other basis you run again the risk that action, the whole purpose of which is military preparation, can be described as peaceful simply because the final logical conclusion of the action has not yet taken place. Clearly that was not what the General Assembly of the United Nations meant when it spoke of reserving outer space exclusively for peaceful use.

I will not, Mr. Chairman, prolong this discussion any further. Let me conclude, as I concluded this morning, with something which is not more than a personal plea. I do not suggest that you attempt at this stage to give it the character of an agreed conclusion. As a personal plea I belive that it remains valid. Why cannot we take now action of the type which I envisaged this morning? One of the speakers said that I knew why we cannot do so now. I do not know why we cannot do so now. I know perhaps why we do not do so now. But, so far from knowing why we cannot do so now, for my part I believe that we can and must do so, and do so now. Thank you, Mr. Chairman.

Chairman: Thank you very much indeed, Mr. Jenks. I have then to ask you, Ladies and Gentlemen, if you agree with Mr. Jenks on this question of a resolution. Do you agree? Yes (confirmed). May I add that we all hope that the working groups will work rapidly and successfully during the next year.

International Control of Outer Space — Some Preliminary Problems

John Cobb Cooper

Before any plan for international control of outer space can be adopted, preliminary problems must be settled.

I. What is meant by "outer-space"? Obviously it is a geographic area, otherwise is would not be subject to "control", and it must be an area outside the territorial sovereignty of any State, otherwise it could not be subject to *international* control. Under the Chicago Convention, 1944, accepted international law, and national statutes, the "airspace" above the lands and waters of the sovereign State is legally part of the territory of that State. "Outer space" is an area above and beyond the airspace. Therefore some definition of its lower boundary must be determined before any scheme for international control can be made effective.

The United Nations *Ad Hoc* Committee on the Peaceful Uses of Outer Space noted that the upper limit of the airspace and the lower limit of outer space do not necessarily coincide. I propose that the lower outer space boundary be fixed, for the purposes of international control, at the upper limit of the "effective atmosphere", that is to say, at a point where a satellite may be put in orbit, somewhere between 80 and 100 miles above the earth's surface.

II. What are the present rights of self-defense of sovereign States? Consideration (a) of Elihu Root's dictum, (b) of Chief Justice Marshall's judgment in Church v. Hubbart, and (c) of the rights of self-defense under the United Nations Charter, particularly Article 51.

III. Is international control of outer-space inconsistent with provisions of the Chicago Convention? Consideration of (a) Article 3 as to State aircraft; (b) Article 8 prohibiting flight of pilotless aircraft; (c) Article 12 dealing with flight over the high seas.

As part of any final decision on methods for the control of outer-space flight, certain preliminary questions must be answered. This memorandum seeks to raise the most important.

I. What is Meant by Outer-Space?

"Outer-space," for political and legal purposes, must be a finite geographic area, otherwise it could not be subject to "control". Also, it must be outside the areas subject to territorial sovereignty, otherwise it could not be subject to international control. It is therefore suggested that in any international convention for control of outer-space or of flight into such areas the following opening statements be included:

"1. Outer-space, for the purposes of this convention, is defined as the area whose upper or outer boundary is the outer limits of the solar system, and whose lower or inner boundary is the lowest altitude above the earth's surface at which an artificial satellite may be put in orbit around the earth."

"2. The contracting States hereby declare that no State has or can have sovereignty over such area or any part thereof or any celestial bodies therein."

My friends will recall that in 1956, before Sputnik I was launched, I suggested to the American Society of International Law a new convention: first reaffirming absolute sovereignty of the subjacent State up to the height at which "aircraft" could be operated; then further extending limited sovereignty upward to 300 miles above the earth's surface; then accepting the principle that all space above should be free for passage.

In 1957, a few weeks after Sputnik I was launched, I published a memorandum in which I pointed out that my 1956 suggestion was based on earlier and then widely accepted scientific opinion to the effect that somewhere not far below 300 miles the atmosphere had sufficient density to prevent free satellite flight, but that Sputnik I had proved this premise to be unsound. It is nevertheless interesting to note that of the satellites now in orbit around the earth the most recent figures indicate that only those satellites with a minimum altitude of over 320 miles above the earth's surface have an estimated orbital life of more than a few months or a few years.

The legal necessity of fixing the lower boundary of outer-space is clear. Under accepted principles of customary international law, the Chicago Convention, and many national statutes, the area called "airspace" above national lands and waters is dealt with as part of the territory of the subjacent State, which alone has the right to control all flight in this area and to prohibit the entry of foreign flight instrumentalities. However the legal status of outer-space is, or under the proposed convention, would be diametrically opposed. For outer-space we should accept the high seas principles cogently stated by Mr. Justice Storey in 1826 (*"The Marianna Flora"*, 11 Wheat. 1):

> "Upon the ocean, then, in time of peace, all possess an entire equality. It is the common highway of all, appropriated to the use of all; and no one can vindicate to himself a superior prerogative there. Every ship sails there with the unquestionable right of pursuing her own lawful business without interruption; but whatever may be that business, she is bound to pursue it in such a manner as not to violate the rights of others."

If this be the status of outer-space, States could without question delegate to an international body the right to control the use of the area, or could between themselves agree on the type of flight permissible in the area. But whether the proposed control be political or functional, the area affected must be defined. An approximate lower boundary of outer-space must be fixed. As the United Nations *Ad Hoc* Committee on the Peaceful Uses of Outer Space pointed out in its 1959 report, the upper limit of the airspace and the lower limit of outer-space do not necessarily coincide. In fixing the lower limit of outer-space at the altitude where an earth satellite may be put in orbit, no decision is required as to whether the absolute airspace sovereignty of the subjacent State extends upward to that line.

The boundary here suggested would appear to be in the area 80 to 100 miles above the earth's surface. At least one satellite has been put in orbit around the earth at a minimum orbital altitude (perigee) of 99 miles. At an altitude of 70 to 75 miles meteors have been observed in an incandescent, or burning state. At 100 miles altitude, and in fact much lower, the "air" no longer exists. At the earth's surface it consists of about 78 percent molecular nitrogen, 20 percent molecular oxygen, and small quantities of argon, carbon dioxide, and water vapor.

At 50 miles altitude the temperature has dropped sharply, the atmospheric density is only about *one-millionth (1/1 000 000) of the surface density,* and not sufficient to contribute in any degree to the aerodynamic lift of flight instrumentalities.

At 100 miles altitude the temperature has increased up to 2000°F or more, the atmospheric density has further decreased to *one-billionth (1/1 000 000 000) of the surface density,* the oxygen molecules have already broken down into separate oxygen atoms, and satellite flight has been proven practical. This extremely thin gaseous combination of a few nitrogen molecules, oxygen atoms, and perhaps particles of other gases, has little if any resemblance to the substance ordinarily called "air" which we breathe and which is needed to support the flight of "aircraft" envisaged when the Paris and Chicago Conventions accepted the principle of "airspace" sovereignty.

This suggested lower boundary of outer-space has therefore practical as well as legal advantages. It is a real boundary. Whatever future observations prove to be the lowest altitude of free satellite flight around the earth will be the lower boundary of the area subject to international control.

It is my recollection that this general boundary location was suggested at a meeting under the auspices of The American Academy of Arts and Sciences in which I participated. In any event, the proposal did not originate with me. My only contribution has been to analyze and elaborate its possible application.

(Note: The statements as to the composition of the air and figures as to atmospheric density are paraphrased from testimony of Dr. Homer E. Newell, Jr., Assistant Director for Space Sciences, NASA, April 8, 1959, before a sub-committee of the United States Senate Committee on Aeronautical and Space Sciences, Page 127, Part I, Printed Record, Hearings on Senate Bill 1582.)

II. What are the Present Rights of Self-Defense of Sovereign States?

Assuming that the legal status of outer-space, as already defined in this paper, is now analogous to that of the high seas, as I personally believe, or that such status will eventually be accepted by formal international agreement, certain presently existing rights of individual States must be acknowledged and preserved. These are particularly the rights of self-protection and self-defense.

The very limited scope of this memorandum will not permit more than the barest outline of how United States thinking has historically dealt with these questions.

As early as 1804 the noted Chief Justice Marshall said (Church v. Hubbart, 2 Cranch 187): "The authority of a nation within its own territory is absolute and exclusive.....But its power to secure itself from injury may certainly be exercised beyond the limits of its territory." This statement still stands in our jurisprudence. It would be directly applicable to the right of a subjacent State to "secure itself from injury" in outer-space beyond its territorial airspace.

Our introductory lecturer today, Dr. Jenks, in his "The Common Law of Mankind," has quoted from a statement of Daniel Webster, as Secretary of State, in the "Caroline" case (1837) that necessity justifying acts of self-defense (outside national territory) is "confined to cases in which the necessity of that self-defense is instant, overwhelming, and leaving no choice of means, and no moment for deliberation." As Dr. Jenks points out, this statement was cited with approval in the Nuremberg cases. But certainly no emergency could leave less chance for deliberation than a threat from outer-space.

Perhaps the most important analysis of the problem made in the United States is found in an address delivered in 1914 by the late Elihu Root as President of the American Society of International Law. Mr. Root served as Secretary of State and also as United States Senator from New York. In discussing the "right of self-protection" as "a right recognized by international law" he said (8 AJIL—6): "The right is a necessary corollary of independent sovereignty. It is well understood that the exercise of the right of self-protection may and frequently does extend in its effect beyond the limits of the territorial jurisdiction of the State exercising it;" and later, in a much quoted phrase he asserted "the right of every sovereign State to protect itself by preventing a condition of affairs in which it will be too late to protect itself."

As I said in 1959 in a paper on Flight-Space Law published in *Handbuch der Astronautik:* "This principle has already been applied toward the regulation of flight. In 1950 the United States and Canada established air defense identification zones around parts of their respective shores. Admittedly the airspace over the high seas is not territorial space and enjoys the same international status as the high seas themselves. Yet the United States and Canada did not hesitate to establish regulations to prevent unidentified aircraft approaching their shores from the seas. The United States regulation, for example, requires that foreign aircraft must report their presence and identification when not less than one hour or more than two hours average cruising distance via the most direct route to the shore. This is a clear application of the right of self-preservation and self-defense applicable outside national territory and within international flight-space. It may well be that the same right exists for subjacent States to act in outerspace above national territorial airspace to the extent deemed necessary for the protection and defense of the lands below."

In any future agreement for international control of outer-space these national rights of selfprotection and self-defense should be preserved. It is submitted that nothing in the United Nations charter is opposed to this view. While Article 51 deals solely with the right of individual or collective self-defense "if an armed attack occurs against a member of the United Nations," it is my firm belief that this does not take away already existing international law rights of self-protection which have long been supported as part of the international law applicable to all States.

III. Is International Control of Outer-Space Inconsistent with Provisions of the Chicago Convention?

In any convention dealing with the international control of outer-space, the most careful consideration must be given to possible conflicts with the 1944 Chicago Convention on International Civil Aviation. This will be particularly true if the International Civil Aviation Organization (ICAO) is not chosen as the agency to exercise outer-space control.

One of the major difficulties is that the Convention deals solely with the regulation of the flight of "aircraft." The term "aircraft" is not there defined. However, ICAO has adopted an Annex to the Convention containing the following definition: "Aircraft shall comprise all apparatus or contrivances which can derive support in the atmosphere from reactions of the air." Certainly this definition does not include rockets, guided missiles, earth satellites or more fully developed later types of space craft. Ordinarily these do not derive support from reactions of the air. A difficulty, however, will arise in case the ICAO definition is modified or amplified to include flight instrumentalities as aircraft even though they do not derive support during flight from reactions of the air. If that is done the most careful consideration must be given to a possible conflict in the regulatory powers of ICAO and the new space organization, particularly as to which agency can regulate space craft flight while passing through the "airspace" and before entering or after leaving outer-space.

Article 3 of the Chicago Convention specifically prohibits "State Aircraft" from flying over the territory of another State without special authorization. Any new convention for the control of outer-space must make it clear that this provision applies only to flight through the airspace. Otherwise serious difficulties may arise as to the flight in outer-space of those flight instrumentalities which can derive support from the air while operating in the airspace but can proceed beyond into outer-space and continue in flight as space craft.

Article 8 of the Chicago Convention provides that no aircraft capable of being flown without a pilot shall be flown without a pilot over the territory of a contracting State without special authorization. Again this provision may create difficulty if ICAO modifies its definition of "aircraft". Accordingly either by amendment of the Chicago Convention or by any new Convention it should be made clear that this article applies solely to flight in the "airspace". As it now reads there is no height limit as to its applicability.

Article 12 of the Convention deals with the "Rules of the Air". Each contracting State undertakes to keep its own regulations uniform to the greatest possible extent with those established from time to time under the Convention. It then states: "Over the high seas, the rules in force shall be those established under this Convention." As chairman of the Drafting Committee at the 1944 Chicago Conference which brought this provision before the Conference I then understood it to mean, and still do, that States who are parties to the Chicago Convention have delegated to ICAO the power to adopt mandatory flight rules applicable to the flight of aircraft over the high seas. It will be noted that the Convention places no height limit over the high seas as to the applicability of such rules. If ICAO modifies its definition of aircraft so as to apply to space craft, difficulties may arise as to whether the ICAO rules or the rules of an international space agency will apply to flight of space craft over the high seas. Even if the article is construed, as I believe, to be applicable only to flight in the airspace, conflicts of jurisdiction may be present so far as the movements of space craft are concerned in the airspace over the high seas, particularly if the ICAO rules and the rules of the new space agency are not precisely the same. Even then the question of which agency will have power to enforce the rules over the high seas will require very accurate thinking and drafting.

These appear to be the major presently foreseeable difficulties created by international control of outer-space if that control is not vested in ICAO. If it is so placed, the difficulties and possible conflicts will be very greatly lessened. These are questions which must be determined preliminary to any final decision as to methods of control of outer-space and flight instrumentalities proceeding from the earth's surface into outer-space and thereafter returning through the airspace.

Sovereignty over Space

William A. Hyman

What is space? Where does it begin? The U-2 and RB-47 incidents emphasize the need for an international code which will define, and establish lines of demarcation between, air space and outer space and agree upon the use thereof. The lack of such a code, coupled with the claim of each nation of sovereignty over the space above its territory is a principal threat to peace.

The increasing use of air space and outer space is the avenue to either peace or war. The hazards of such use can be overcome by the cooperation of scientists, lawyers and diplomats. Scientists must inform the public of all scientific advances. Lawyers must devote themselves to the drafting of laws, obedience to which will be compelled by informed public opinion. Politicians must cooperate and forswear arbitrary action.

The satellites launched into orbit around the earth, the moon and the sun, and the 8000 mile rocket shots attest to science's technological progress. Unfortunately, there has been no progress in international law since the 1959 Report to the United Nations by the *Ad Hoc* Committee on the Peaceful Use of Space.

There is sufficient scientific and legal knowledge to permit creation of an international code, delineating rights and obligations of all nations. Agreement on the areas of air space and outer space, and that outer space is *res communis* and *not terra nullius,* proscribing planetary claims, is feasible. Without it there will be no peace, but always the threat of annihilation.

No doubt certain differences exist between Communism on the one hand, and Capitalism under Democracy, on the other hand, but these differences are being narrowed through the process of time, patience and the voice of the people. There is no necessity to resolve these differences by combat which would be catastrophic to the world but they must be and can be determined by the adoption of a fundamental skeletal outline acceptable to all peoples. This is true even if the adoption of such a skeletal outline means a sacrifice of some sovereignty on the part of each nation while undergoing the transition from a "nation-state" to a "world-state".

This procedure will prevent provocations for international tensions and war. It will establish a system of law and order rather than chaos in Outer Space. This is most desirable since it is recognized that space is the most vulnerable spot in the armor of every nation today.

In law and order this is justice! In justice there is peace!

There is void in international law. Rocketry has been too fast for global law.

If a Russian test missile veers off course and crashes into the heart of Stockholm, who is liable to the injured persons and to the widows and orphans of the dead for damages?

If an American missile collides with an S.A.S. airplane loaded with passengers and all are killed, who will pay damages? and how?

If two satellites collide in orbit, and fragmentation passing through the radiation belt strikes and destroys the Empire State Building and injures and kills crowds of people, is there any legal redress? and against whom?

Sovereignty over Space

With the stepped-up activity in space — the United States alone having announced a ten year program of two hundred and sixty shots — and other nations also embarking on ambitious programs — accidents and mishaps will occur bringing injury and death to helpless innocent persons and tremendous damage to property.

Can these victims recover? They cannot. There is no law today giving them an absolute legal redress against any offending sovereign party.

And what will be the remedy of the sovereign nation whose nationals are the victims? In law — none.

Since October 4, 1957, when Sputnik I heralded the advent of the Space Age, activity in space has been greatly accelerated. This demands a prompt revision of attitude, and immediate reappraisal by all nations.

The progress of science in the new world of space has been sensationally rapid. The progress of law therein has been slow. As to diplomacy — the field of international politics — not only has there been no progress but there has been retrogression which may threaten the peace of the world. Originally, it had been hoped that the scientist, the lawyer and the diplomat, working together, would help avoid conflict by bringing forth certain answers to these problems in the form of a skeletal outline of an international code for regulating the peaceful use of outer space.

Recent events indicate that the burden for preserving peace rests upon the scientist and the lawyer and not so much upon the politician who has failed signally to achieve international understanding and who has caused a serious threat to world peace.

The exploration of space, swift and awesome, must force all thinking people to a re-evaluation of the entire situation and to immediate affirmative steps to avoid war and the possible total destruction of mankind.

I.

Let me note a few of the sensational developments in science.

Polaris missiles fired from submarines beneath the surface of the sea strike a target almost 1200 miles distant.

The development of the first true amplification of light might soon lead to the capacity to generate a fine light beam which could develop sufficient intensity to illuminate parts of the moon's surface and to vaporize materials in its path, notwithstanding that the moon is over 280 000 miles from earth. It could also be used to devastate parts of earth.

In January and again in July of 1960, rockets were launched a distance of over 8000 miles into a target area in the Pacific presumably as a preparation for a rocket shot to Mars or Venus. In like manner, the United States launched the Atlas intercontinental ballistic missile a distance of 9000 miles into the Indian Ocean on May 20, 1960.

Incidentally, it is noteworthy that the January rocket of the Soviet landed at a point approximately 500 miles from Johnston Island in the Pacific, one of the most important naval installations of the United States.

New magnetic laboratories are being installed to probe the magnetic field in outer space. This will help science to measure the influence of powerful magnetic fields in such tiny particles as electrons, protons and neutrons and lead to the discovery of new surprises in the cosmic system bringing to light the vast reaches of outer space including the bombardment of earth by cosmic rays.

The National Aeronautics and Space Administration (NASA) of the United States in two years has achieved some brilliant successes. Tiros I, the weather satellite, equipped with television and sending back over 20 000 photographs in a week, and Pioneer V, which radioed scientific information back to earth across millions of miles attest to its great success.

The Soviet and the United States both are making plans to launch manned space craft in the course of the next few years. Dogs and monkeys have already been launched into orbit with safety and brought back without harm.

Rockets as message links and as a force in a system of world communication are now in process of development. This space communication system may be an absolute necessity in the future if other systems are impaired or destroyed.

Credit must be given, of course, to the Soviet which successfully launched Sputnik I on October 4, 1957, followed by Lunik I in January, 1959, which went past the moon and into orbit around the sun and for Lunik II which on September 13, 1959, hit the moon, at which time, allegedly, their flag was dropped on the moon. Credit must be given to their scientists for the rocket sent into orbit around the moon and then directed in a path into orbit around the earth on October 3, 1959. These reveal the excellence of their guidance system and their accumulation of vast knowledge in this new science.

The conclusion is now inescapable that both East and West have made tremendous progress in the scientific development in this new world of space.

In the effort for better understanding necessary for a peaceful solution of urgent world problems, a new contribution has been made in the change of attitude of the scientist towards the public. Originally, the scientist was loath to reveal his findings to the public. Also, at various times he had been subjected to restraint. The American Association for the Advancement of Science, however, rendered a report in July, 1960, in which it asserted "that the overriding public issues of the day concern scientific matters. It is, therefore, essential for the preservation of a democratic society that scientists develop an informed public in such matters". It is likewise now urged that scientists of all the world adopt the same standard of open cooperation with, and education of, the public.

Great has been the achievement of science. Greater yet will be its contribution to peace, law and order in society if the scientist furnishes to the lawyer the answers to some questions.

II.

Before a solution is suggested in what appears to be an almost hopeless situation, it would be desirable to offer some analyses of space, sovereignty and international law.

The *Ad Hoc* Committee of the United Nations reported in June, 1959, that the determination of precise limits for Air Space and Outer Space did not present a legal problem calling for priority consideration at this time. Yet, it seems to me that such a determination is immediately necessary as a basis for agreement on the use of Outer Space and the limits of Air Space. Without such definition there may be conflict concerning what has been agreed to and conflict as to national rights to Air Space as distinguished from Outer Space. The divergent and shifting views of the United States and the Soviet Union as to a nation's complete and unlimited sovereignty over the space above its territory indicates the need for a definition of Outer Space and an agreement on where it begins. The problem of where Outer Space begins may be considered a sequel to the problem, where does the ocean end? This is not as preposterous as it sounds if one refers to recent actions of United States Courts in extending the jurisdiction of the Federal Death on the High Seas Act to wrongful acts in the air over the ocean. In the important case of *D'Aleman v. Pan American World Airways, Inc.* [1] the Court deliberately extended a law intended to apply to acts on the high seas to apply to acts occurring in Air Space over the high seas, for the purposes of providing some remedy for a claimant who otherwise would have had no remedy, because there was no law of air or space to govern such a situation.

How much more complicated will be the situation where a wrongful act or death occurs in Outer Space!

While sovereign powers claim ownership of Air Space, there has been no consensus concerning precise limits for Air Space.

Therefore, it is with great timidity and hesitation that any suggestion is made for any theory at present but it is desirable that some step be taken forward which might simplify as much as possible this very difficult problem.

At present, scientists agree that there are five layers which constitute the atmosphere around the earth. Perhaps by agreement Air Space can be held to comprise the troposphere and stratosphere extending to a point about forty kilometers above the earth's surface and all above that area might be deemed to be outer space. By proper technological means with current inspection of instrumentalities and means of determining course and position, and under proper international police supervision, the adoption of this separation might tend to guide vehicles in transit, objects, satellites, space craft against violations. The arbitrary fixation of a limit at a given distance might prevent confusion and provide a more stable guide than those theories which provide boundaries of a fluctuating nature and conceded instability.

The right of sovereignty by any one nation over a given area necessarily gives to that nation the right to exclude all other nations therefrom. If asserted against the area of space, spaceways and airways, catastrophic results might develop. Serious provocations might ensue and set off incidents leading to a grim end from which there would be no turning back.

The RB-47 plane incident over the Barents Sea alleged by the United States to have occurred from fifty to seventy-five miles off the Soviet Arctic coast on July 1, 1960, the U-2 plane flight incident on May 1, 1960, occuring in a course of action (acquisition of information) which has been pursued by the Soviet and other nations for many years; the presence of Soviet trawlers and vessels at or near the three mile limit off the American coast observing American experiments and maneuvers; the presence of Soviet vessels off the British coast making electronic observations; routine Soviet airflights off the coast of Japan, so frequent as to be called the "Tokyo Express" (to obtain data and information) — all these incidents emphasize the urgency of immediate international agreement controlling the use of Air Space and Outer Space without delay.

The Soviet shot rockets into the central Pacific in January, 1960, and again in July, 1960, after ordering vessels of all nations to remain out of an area estimated about 40 000 square miles in size. Did this constitute an assertion of domination, control and sovereignty over the peaceful commercial travel lanes in Air Space and on the high seas? If the open seas and the airways above are *"res communis"*, then no one nation has the right to exclude other nations therefrom. This right of exclusion is incidental to the right of ownership and sovereignty.

Accordingly, I suggest that the following definitions be considered for insertion in an International Code:

> "Space is that area existing between the surface of the earth and the celestial bodies. It shall comprise two parts. The first part shall comprise the troposphere and the stratosphere. The second shall comprise the remainder of the area extending to the celestial bodies, which shall be termed 'Outer Space'.
>
> "The troposphere shall be deemed to extend from zero to ten kilometers above the earth's surface.

"The stratosphere shall be deemed to extend from ten to forty kilometers above the earth's surface."

The *Ad Hoc* Committee mentions the question of freedom of Outer Space for exploration and use and merely states that the International Geophysical Year (1957/8 and subsequently) may have initiated the recognition or establishment of a generally accepted rule to the effect that, in principle, Outer Space is, on conditions of equality, freely available for exploration and use by all in accordance with existing or future international law or agreements. As to the problem of exploration, exploitation and settlement of celestial bodies the *Ad Hoc* Committee reported that these were not likely in the near future and did not require priority treatment.

However, it will only be through international law that the principle of freedom of exploration and use of Outer Space can be established as acceptable to all nations. Without agreement thereon conflict is inevitable. Moreover, exploitation of Outer Space and even of celestial bodies may not be as far off as the *Ad Hoc* Committee seems to think, and, therefore, it is advisable that agreement be reached that celestial bodies be rendered incapable of appropriation to national sovereignty and that exploration of Outer Space and of celestial bodies be carried out exclusively for the benefit of all nations and all mankind.

Violations of sovereignty always constitute the greatest threat to peace. The establishment of boundaries in territorial waters and on the land have proven a restraint upon trespass of sovereignty although even at times in these fields, incidents have developed because of the innocent crossing of such boundary lines. While in Air Space and in Outer Space it is impossible to fix such lines of demarcation with anything like the stability of those boundaries of earth, of the sea and in territorial waters, the establishment upon some settled principle of such boundaries between Air Space and Outer Space will likewise prove a deterrent to provocation and international friction. Now, more than ever, the world needs the immediate guidance by the scientist in establishing a formula or principle of separation and determination of boundaries. The people of the world demand the settlement of these questions through peaceful negotiation — not by attack, destruction and killing. These peaceful negotiations must be conducted pursuant to international law at once and without delay because of the present state of world tension.

What is meant by "international law".

It has been stated as follows:

"International law consists of certain rules of conduct which modern civilized States regard as being binding on them in their relations with one another with a force comparable in nature and degree to that binding the conscientious person to obey the laws of his country and which they also regard as being enforceable by appropriate means in case of infringement" [2].

But, it has also been said:

"We must expand our interpretation of the term 'international law'. We must cease to think of it as merely a set of principles to be applied by courts of law, and understand that it includes the whole legal organization of international life on the basis of peace and order. Such an organization must provide for peaceful and orderly use of political, as well as judicial, methods of adjustment" [3].

What about "space law"?

It is most regrettable that up to the present time no written international code, even in the most skeletal form, has been adopted by the nations on this subject.

The most important effort to meet this problem was made in the United Nations but that effort has been productive of very little good. A resolution was adopted on the 13th of December, 1958, establishing the *Ad Hoc* Committee on the Peaceful Uses of Outer Space. The report was adopted by the Committee on the 25th of June, 1959, and it was later adopted by the General Assembly on December 12, 1959. A new Committee (a Standing Committee) was formed by resolution of the General Assembly on the 12th of December, 1959. Up to the present time, this Committee has not even met.

The report of this *Ad Hoc* Committee offered no solutions. It merely posed certain questions. It established priority for some and a secondary position for others. It suggested that a comprehensive code was not practicable or desirable at the present time. However, its very capable counsel did suggest that at least a skeletal agreement be established which could be amplified subsequently in accordance with the development and solution of these future problems.

In July, 1960, the Soviet Premier threatened with destruction Austria, a sovereign power, if missiles were launched from Italy which traversed the space above Austrian territory, notwithstanding the fact that Austria would have no basis for control of such passage nor even be in a position to prevent it, even if Austria knew about it. When a threat is made to send rockets from the Soviet to the United States, this implies that such rockets would pass over areas of Outer Space and also Air Space of intervening countries. Would they be held responsible in the same manner in which the Soviet seeks to impose responsibility on Austria for missiles launched from Italy?

The problem is well stated in the National Space Program Report of the Select Committee on Aeronautics and Space Exploration (United States, 1958, pp. 22-23).

> "A major legal problem raised by space flight concerns the upper limit of sovereignty over space above national territory. There are a number of possible solutions, ranging from unlimited national sovereignty upward, through international control and regulation of outer space, to complete freedom of the use of outer space by all nations for all purposes. Unless national sovereignty in outer space is to be unlimited, each of these solutions involves an international limit or definition.
>
> "Existing international agreements refer to sovereignty only in the airspace over national territory and territorial waters, and hence do not apply, in terms, to outer space. As Mr. Becker testified, the United States has never agreed to an upper limit to its own sovereignty. In addition, he argued that satellite flights up to now are sanctioned only by an implied international agreement. This is based on the tacit acquiescence of all governments in the announcements by the United States and the Soviet Union that satellites would be launched in connection with the International Geophysical Year. It is therefore limited to the types of satellites contemplated in those announcements and to the duration of the International Geophysical Year. Mr. Becker's statement to this effect constitutes a major declaration of national policy."

In the unfortunate controversy in this field between the Soviet and the United States, expediency rather than legal status seemed to govern action, as revealed by reversal of positions by both sides.

In 1956 the United States, Great Britain, Turkey, and West Germany were conducting meteorological studies necessitating the flights of unmanned balloons through the stratosphere at a level of 80 000 to 90 000 feet. Some of these balloons flew over Soviet Territory and some even landed on Soviet and Soviet-satellite soil. The Soviet Union and most of its East European allies protested to the United States that it had trespassed on their sovereign territories.

On February 7, 1956, Mr. John Foster Dulles, then Secretary of State of the

United States, maintained that there was no rule of international law on the subject and that flight of one nation's balloons over another nation's territory at the height of 80 000 feet was legal and could not be objected to by the subjacent state, since the question of the ownership of such space was obscure and disputable.

Soviet policy on this point was in direct conflict. Article I of the Air Code of the USSR of August 7, 1935, states that "to the USSR belongs the complete and exclusive sovereignty in the air space above the USSR". This is the *Ad Coelum* theory. The Russian scientists, Koslov and Krylov, stated that this meant that the Soviet sovereignty was *without limit*. Subsequently, however, in September, 1958, the Soviet legal expert, Miss A. Galina, evidently speaking with the approval of the Soviet Government, contended that since there was no international law covering space any government might launch rockets or satellites into interplanetary space without the permission of any other government.

On the other hand in March, 1958, Loftus Becker, then Legal Adviser to the State Department of the United States, stated that the United States could still claim and defend all space above its territory.

III.

How can these conflicts be resolved? The answer is in the creation of an international code even though this would involve a surrender of some rights of sovereignty. It portends inevitably the transition from the nation-state to the world-state for the protection of mankind.

Bertrand Russell recently stated that the world must accept the doctrine of world authority or suffer the extinction of the species. In submitting to international law and becoming a party to an international treaty a nation inevitably suffers some loss of sovereignty.

Will the leading powers submit to an international agreement which implies the creation of a "world state" and curtails some of the powers of sovereignty? Is such an international agreement a means of avoiding deviation by any nation from its expressed principles of mutual consideration and cooperation? In what manner can sovereignty be curtailed for the benefit of world peace without loss of individual status?

Fenwick in his *International Law* states:

> "The extent to which the doctrine of the sovereignty of states operated as a standing obstacle to the development of an organized community of nations cannot be exaggerated. In its extreme form, the doctrine implied the complete freedom of the State from the control of any higher power claiming authority to regulate its acts. It was a doctrine of legal anarchy ..." [4].

The claim of sovereignty, carrying with it ownership and control, has weighed heavily on the world. Today, more than ever, the question is acutely posed because of the dramatic accomplishments in the world of science. What will be the implications if a manned space craft lands on the moon? What will be the implications if, in addition to planting a flag, there is an attempt at establishing a settlement thereon?

Although the Soviet stated on the occasion of Lunik II striking the moon that no claim would be made of ownership of the moon, nevertheless, it is conceivable that later developments might cause a change in this attitude. This makes it necessary to analyze in advance the validity of any such claim.

In days gone by, such an act as planting a flag, a sword or a cross on territory, so as to take symbolic possession, did constitute a basis for a claim of extension of sovereignty. Today, however, territorial sovereignty requires effective occupation, with the right to exclude other states from a region and the duty to display therein the activities of a state. This concept was stated by the eminent Swiss Jurist, Dr. Max Huber, acting as arbitrator in the case of the Island of Palmas, which is located forty-eight miles southeast of Mindanao in the Philippines. The Permanent Court of Arbitration of the Hague held:

> "The growing insistence with which international law, ever since the middle of the 18th century, has demanded that the occupation shall be effective would for the act of acquisition and not equally for the maintenance of the right ...
> "The title of discovery, ... under the most favorable and most extensive interpretation, exists only as an inchoate title, as a claim to establish sovereignty by effective occupation" [5].

There remain on the earth only a few unexplored regions. These are principally the Arctic and Antarctic areas. The United States has consistently opposed all claims of sovereignty over any part of these regions by any nation as a result of discovery. The United States' stand against Norway's claims following Amundsen's explorations was expressed by Secretary of State Hughes writing to A. W. Prescott on May 13, 1924:

> "It is the opinion of the Department that the discovery of lands unknown to civilization, even when coupled with a formal taking of possession, does not support a valid claim of sovereignty unless the discovery is followed by an actual settlement of the discovered country" [6].

To the Norwegian Minister, H. H. Bryn, Hughes wrote on April 2, 1924:

> "In my opinion rights similar to those which in earlier centuries were based upon the acts of a discoverer followed by occupation or settlement consummated at long and uncertain periods thereafter, are not capable of being acquired at the present time. Today, if an explorer is able to ascertain the existence of lands still unknown to civilization, his act of so-called discovery, coupled with a formal taking of possession, would have no significance, save as he might herald the advent of the settler; and where for climatic or other reason actual settlement would be an impossibility, as in the case of the Polar regions, such conduct on his part would afford frail support for a reasonable claim" [7].

The Soviet Union has indicated agreement with the United States that national territorial claims should not be recognized unless a nation can effectively occupy the area claimed. Nevertheless, seven nations have claimed sovereign territorial rights to slices of Antarctica. On December 1, 1959, a treaty was signed by twelve nations, including the United States and the Soviet Union "freezing" the national territorial claims to Antarctica. This Antarctic Treaty provides that the claims by the seven nations are not to be affected in any way by the Treaty, but that no new claims can be made while the Treaty is in force [8].

This Treaty is a great historical event. It indicates the ability of nations to subordinate conflicting claims of national sovereignty to international cooperation for the benefit of all mankind. It is a splendid precedent for a similar treaty with regard to Outer Space.

Before a crisis is reached with regard to the moon or any other planet or Outer Space, by possible adverse claims of sovereignty, immediate agreement should be reached on an International Code prohibiting claims of sovereignty on the part of any one nation and establishing rules for internationalization of Outer Space and the Interplanetary System.

IV.

It has been pointed out that there is an immediate need to establish an appropriate international code to settle some of the primary and elementary questions. It has been shown that the chief source of provocation might lie in offense to sovereignty created by the failure to determine the boundary lines separating the areas of sovereign control and that area which is not subject to such control.

The solution lies in joint action of all bodies of law, science, and politics joining to force action at the United Nations and its Committee to regulate the use of Outer Space even though its first product will be not a complete code but a skeletal outline, setting forth certain basic provisions which will avoid confusion and conflict.

In this connection, it is suggested that such a skeletal outline of international convention provide as follows:

a) That all space be divided into Air Space and Outer Space;

b) That all Air Space be deemed to be part of the sovereign jurisdiction of the subjacent land;

c) That all Outer Space be deemed *res communis* (and not *terra nullius*);

d) That the interplanetary system be deemed *res communis* (and not *res nullius*);

e) That recognition be given to the distinction between *"res communis"* and *"terra nullius"* (the former denying rights of appropriation and exclusive control by any one nation, the latter conceding such rights of appropriation through the established principles of discovery, habitation and settlement);

f) Furthermore, since it is impossible to set out a boundary line with physical qualities such as characterize boundary lines on land and on the sea, that there be established a neutral zone between the upper limits of Air Space and the lower limits of Outer Space to be known as "Neutralia" in which the right of innocent passage shall be recognized without offense to sovereignty. In this area the vehicle and/or person in transit shall be entitled to warning and guidance without being subjected to attack or destruction;

g) That the rights and obligations of the nations of the world in and to each of the aforesaid areas should be set forth and the exercise thereof be determined through negotiation and arbitration and not by combat;

h) That police supervision for operations in space be provided, the means to do this to be evolved by qualified scientists.

i) That provision be made to establish an international insurance fund to indemnify all persons for damages to person, life and property caused by falling missiles, fragmentation from satellites, irradiation, foreign forces from Outer Space, and other relevant operations in space.

Of course, no attempt can be made here and now to seth forth details of this plan and procedure. These will have to await the acceptance of the principle.

Conclusion

Peace is the avowed desire of all people and all nations. The Communist countries have proclaimed their support of the doctrine of peaceful coexistence. The West likewise has indicated support of not merely peaceful but continued existence with all nations permitted to select their form of government and their own ideology and that this selection be determined through peaceful means.

Time and patience and not combat will provide the final solutions — by the will of the people.

This appears all the more probable because certain differences existing between the two ideologies have appeared to have been narrowed.

According to the press, Professor Leonid V. Kantorovich, the Soviet Union's leading mathematical economist has proposed certain reformations that amount to abandonment in part, if not entirely, of the labor theory of value formulated by Karl Marx [9]. The capitalism which existed in the days of Karl Marx with its monopolies and exploitation, no longer exists. The transition has been definitely away from the extreme right towards the partial adoption of paternalistic principles and even towards some pseudo-socialistic views.

Senator Henry Cabot Lodge in a speech delivered before the Economics Club in New York on September 17, 1959, said:

> "We live in a welfare state which seeks to put a floor below which no one sinks but builds no ceiling to prevent man from rising."

This is a far cry from the old capitalism which was the subject of so much socialistic attack. The merits of each system in the controversy between East and West must be decided in a peaceful competition and not by a war of annihilation; by law, expressing the will of the people and not by combat. The obstacle to this accomplishment is the doubt concerning the sincerity of the proclamation and conduct of spokesmen for each side. It is hard to believe that there is true friendship on the part of one who shakes his fist under the nose of another threatening the other with destruction if he does not accept his views. The hurling of insults and humiliating threats do not create a spirit of confidence and the healthy climate necessary for a peaceful solution.

The adoption of an international agreement even in a skeletal outline as has been suggested herein and also by the distinguished Committee on the Peaceful Uses of Outer Space of the United Nations, would go a long way towards eliminating the basis for unpleasant provocative conduct which has characterized the present period of tension.

Space is the most vulnerable spot in the armour of every nation today because the control of space would enable an aggressor to devastate the earth. The nation which controls space will control the world. If this control falls into the hands of any ruthless, dictatorial nation governed by unscrupulous and shortsighted politics, it may mark the end of freedom for mankind. It may even mark the end of mankind. An international space law provides the means for anticipatory control as well as present regulation and thus can avoid such a dreadful end. Action must now supplant words. The law of space cannot wait for slow development over a long period of time as did the law of the horse and buggy, the automobile, the ship and the train. Science is moving too rapidly for such slow procedures.

Outer Space, now being explored by individual nations will undoubtedly soon be put to world use. Shall we sit idly by? Shall we do what we can to insure that the use of Outer Space be devoted to the common benefit of all mankind? With the co-ordinated teamwork among the scientist, the lawyer and the politician we can achieve this service to mankind, the elusive objective — peace. In this manner shall we be able to guarantee not merely coexistence but the very existence of all nations big and little.

In lawlessness there is chaos and perhaps extinction. In law there is survival, order and peace. In peace there is justice.

References
[1] 259 F. 2d 493; October 2, 1958.
[2] *Hall,* International Law, (8th Ed. 1924) p. 1.
[3] *Brierly,* The Rule of Law in International Society (1936) quoted in Bishop, International Law (1953) p. 2.
[4] *Fenwick,* International Law, (3rd Ed. 1948) p. 29.
[5] Permanent Court of Arbitration, 1928, Scott, Hague Court Reports 2nd ser., p. 83 (1932); 2 UN Reports Arb. Awards 829.
[6] *I. Hackworth,* International Law, 399.
[7] *Ibid.* 453.
[8] New York Times, December 2, 1959, p. 1, col. 5, p. 46, cols. 1—8.
[9] New York Times, June 12, 1960, Sec. I, p. 10.

Survey of Legal Opinion on Extraterrestrial Jurisdiction

Andrew G. Haley

This survey covers those aspects of international control of outer space relating to the questions
1) Where does the airspace jurisdiction of terrestrial states end under existing treaties and under the internal statutes of the nations of the earth?
2) What opinions have been asserted concerning the rule of law in extraterrestrial space?
Great confusion is found to exist in the writings of commentators because of the mingling of authentic legal postulates with speculative political considerations. The bulk of commentary is thus hopelessly devoid of distinction between the dichotomy of law and politics. The jurist is misled by the careless, ambiguous and cursory use of the terms "atmosphere" and "airspace" by the natural scientists — engineers and physicists. In law these terms have definite meaning, and the meaning, as applied to particular cases, solves the venue of applicable crimes, torts, civil liability, criminal liability, and the rights, obligations, and responsibilities of men and nations.

When the political coating is removed from the heterogeneous mass of commentary, it is found that remarkable agreement exists between the writings of the Soviet and Peoples Republics experts and the viewpoints of their brother legal experts in other countries of the world.

In arriving at any considered opinion it is necessary first to determine actually the terminal point of terrestrial jurisdiction as legally, not politically, defined, in order to consider the problems of the rule of law in extraterrestrial space. It is found that the weight of authority favors the adoption of the "von Kármán line" as a terminal point of terrestrial jurisdiction for civil and criminal venue, and the problems of establishing the rule of law in outer space are adverted to briefly.

When I was President of Aerojet, I suddenly was confronted with a most difficult military requirement, which necessitated the employment of several first-class physicists. I initially interviewed a truly great physicist — he now has a "household name", and I told him of my problems. Towards the end of the conversation, I also told him that, as an outgrowth of our work, we sincerely believed that man would eventually conquer space. The professor looked at me quizzically and said "Ah, this is impossible as your exhaust gases will have nothing to push against in outer space". I remained silent for a few moments. Then I said "Good Professor, I must now answer the telephone. Would you please be kind enough to permit me to leave you, and when I return will you discuss with me Newton's Third Law of Motion?" I stayed away purposely for one-half hour, and when I returned this great man was smilingly apologetic and disconcerted. It was obvious that he had gone back to his days in the Gymnasium and had recalled the long unused Newtonian Laws. That was the beginning of a fine and enduring friendship and between us we have never mentioned the incident since.

In the same strain and in connection with Mr. Phillip W. Quigg's article in *Foreign Affairs*, October 1958 — "Open Skies and Open Space" — and especially with reference at Page 99, I had hoped that the charity of my silence on the wholly unscientific quotation which follows — a silence which was induced by my immense regard for Mr. Quigg — would eventually eliminate the necessity for an answer [1]. — Here is the quotation from Quigg's article:

"The clarity of Haley's proposed boundary has been muddled considerably by the announcement that the Air Force plans to fly an experimental aircraft to more than twice the altitude at which aerodynamic lift is gone. The X-15 will thus soar well above the perigee of Explorer III (110 miles) and may reach that of the first three sputniks (145 to 150 miles). It will accomplish this, however, much as an cold car gets over a hill — by gaining maximum speed on the straightaway and counting on its weight and speed to carry it up the grade. So the X-15 is expected to achieve a speed of some 5000 miles an hour in level flight with most of its weight supported by aerodynamic lift; then turning upward, and with a final booster from its rocket motor, it will coast into the realm of satellites, gliding back to earth immediately, its energy exhausted.

"The significance of this vehicle is that it blurs the distinction between aircraft and spacecraft and may indeed be the prototype of future space ships. The first human to be placed in orbit is likely to be borne in a winged craft, not a pure rocket. The wings will give him greater stability and control in take-off and ease his glide back to earth. And before space stations are available to be used as transfer points from aircraft to spacecraft, the descendants of the X-15 may be so highly perfected that flight will be an unbroken spectrum from atmosphere to space. While this does not invalidate the distinction between zones of aeronautical and astronautical flight (even Under Secretary of Defense Quarles has supported the concept as a way of distinguishing between air space and outer space), it becomes somewhat more doubtful that the boundary proposed by Haley will be accepted as the limit of national sovereignty."

Much to my annoyance, and the annoyance of those who have the slightest knowledge of the scientific facts, the foregoing unblemished sophistry and "scientific" nonsense has been repeated time after time and has even achieved the dignity of an illustrative reference parameter in many articles [2]. And one is baffled in attempting to answer or explain — as the statement is so irrational and unrelated to facts — any facts — let alone the Kármán jurisdictional line. Even the publicists are led into calling the X-15 an "aircraft", whereas it doesn't breathe any air and is in fact a pure rocket vehicle. The X-15 may function equally well in or outside of the air but the maneuvering must depend on apposite propulsion and guidance principles.

Mr. Quigg states "The first human to be placed in orbit is likely to be borne in a winged craft, not a pure rocket". If Mr. Quigg had been referring to a propulsion system such as an "athodyd" or "ramjet" (technically an aerodynamic, thermodynamic duct) which would be used during the flight through air and later on with a second rocket stage which would become operational after the ramjet had passed through the air [3], and if at the same time he were referring to frictional guidance during the passage of the vehicle through the air — then at least the description of his vehicle might make some sense, but such description cannot be related to "winged craft, not a pure rocket" and indeed to any aspect of the concept of the Kármán line [4]. So the thoughts expressed in the sentence are hopelessly confused, and the confusion cannot be converted into good sense.

The whole excerpt becomes still more inexplicable in view of the illustration he gives, namely the X-15. The X-15 is a pure rocket. It has no air-breathing devices and depends for its own propulsive capacity on pure rocket power only. The configuration of this vehicle permits the temporary use of air for guidance. Once this vehicle leaves the earth's air, it must depend upon the trajectory already established or upon some other method of guidance. If the X-15 is intended solely for use in the air, then it is an airplane using rocket power. In all justice to Mr. Quigg it seems even some NASA officials do not understand the basic principles [5]. Under the preposterous assumption of the above quotation, however any vehicles — vehicles designed for earth orbiting or flights to the moon, the planets or indeed to the stars — if they used air guidance surfaces during the brief seconds they depart from the earth or return to the earth, they would not be space vehicles. Indeed, if they utilize the earth's air in any fashion, they would not be space vehicles. Thus, the use of a parachute or any form of air drag brakes would, according to Quigg, mysteriously transform Sputnik V, for example, from an earth orbiting satellite to an aircraft and this transformation would subject Sputnik V to the airspace jurisdiction of all of the nations of the earth. Along with the X-15, the Dyna-Soar would become an aircraft because this space vehicle, which is a boost-glide orbital space craft, depends on air glide on departure and upon return and is proposed to be fitted with large stabilizer fins on the booster [6].

The Mercury, designed as a manned satellite also would be a mere airship, as the two-stage Mercury landing package uses improvements of existing equipment, aimed at a parachute landing system (air brakes, of course) with maximum reliability. Mercury landing package consists of a drogue parachute, main and reserve recovery chutes, aneroid and inertial sensors, explosive actuators and recovery aids. Normal landing sequence begins at 42 000 ft following capsule re-entry from orbit. The aneroid instrument senses this altitude and closes an electrical circuit to ignite a mortar shell, ejecting the 6 ft ribbon drogue parachute. The drogue chute, first stage in the system, rapidly inflates to stabilize the capsule and to slow it from Mach 1 velocity to a true airspeed of 160 kt [7].

Even the epochal Apollo, three-manned capsule for lunar orbits and other space missions, would also be designated as an airship, because this space vehicle most probably will have aerodynamic glide capabilities and will be fitted with parachute brakes.

From descriptions which will follow and particularly in the light of the information contained in "Air Space Jurisdiction — the Statutory Law of the Nations of the World" [8], the present state of international law appears to be (1) that each nation is sovereign in the "air space" over its land mass and territorial waters; (2) that the upper limit of sovereignty actually ends where air ends; (3) that under well known principles a nation is justified in protecting itself in air space, on the high seas, or in outer space, or anywhere else, if attacks or belligerent actions are being staged in any such areas; (4) that passage through and the use of outer space is completely free from any restrictions, if the passage or use is for peaceful purposes [9].

I read with great interest an emerging body of commentary which would confine national sovereignty to the upper limits of the area in which a nation may defend itself — this being some sort of a concept of "effective control". I have also studied commentary which purports to excuse a violation of air space on the grounds that inspection by this means is essential for the continued "peace of mind" of the peoples of the world. These speculations are well beyond the scope of my position which is founded strictly upon existing international and municipal law. The philosophy of the new commentary may at some point prevail. At the present time, however, taking the instance of my own nation, the United States of America, I would unhesitatingly invoke the sanctions of the municipal laws of the United States, if such laws were broken by any foreign aircraft for any reason whatsoever. Let me repeat again, I refer to *aircraft* flying in the medium of *air* — and that is all to which I refer.

As has been adverted to above, and which will be discussed later in more detail, more than 70 nations, speaking individually through their constitutions or statutes, have asserted their sovereignty over the airspace above them. The International Civil Aviation Convention adopted at Chicago in 1944 is the most recent multi-lateral expression on the subject. The agreement reached then, like the predecessor Paris Convention of 1919, is couched in terms of aircraft, defined in Annexes 6, 7 and 8 as "all machines which can derive support in the *atmosphere* from reaction of the *air*". (Emphasis supplied.)

Sixteen years later, we live in another world, technologically speaking, for in the space that we probe today, there is (as we have seen) no *air* or *atmosphere* [10].

To repeat, when we enter the regime of rockets and satellites, we rapidly reach a point beyond which that "gaseous substance" thins out, *i.e.*, its molecular oxygen dissociates into atomic oxygen [11], as do the other components.

Where exactly is that point which separates airspace from what is variously (and loosely) called outer space or cosmic space, and more important, which separates the realm of national sovereignties from a domain of international space law in the making? If we can arrive at a workable definition of this point (or more accurately, line, for it is an irregular and varying boundary), a long step will have been taken toward clearing away the semantic underbrush from the site on which must be erected the structure of international space law.

It is neither possible nor necessary to lay down an absolutely rigid line between airspace and "outer" space. Such a concept partakes of pure casuistry. In all science one must deal with a median based upon an immense family of curves. As a practical guide for the space age, the weight of authority favors a measure of the sort I have termed the Kármán primary jurisdiction line [12]. This line was suggested by Dr. Theodore von Kármán and adopted by the writer on the basis of a diagram by Masson and Gazley of the Rand Corporation. An adaption of this diagram is attached hereto as Appendix 5. Simply stated, the Kármán jurisdictional boundary falls at approximately 275 000 feet (83 km), where an object traveling at 25 000 feet (7 km) per second loses its aerodynamic lift and centrifugal force takes over.

The line may be changed somewhat as physicists and lawyers hammer out agreement as to where the aeronautical vehicle no longer may perform efficiently and within reasonable physiological and engineering parameters. It may be useful to glance briefly at some of these parameters:

A. M. Mayo (in *op. cit.*, Note 29) points up a number of basic environmental problems. At altitudes above 70 000 feet (21 km), he notes, pressurization of outside air becomes increasingly difficult in terms of power required and the very high temperatures resulting from extreme ratios. In addition, the threat of decompression will be acute until pressurized cockpits are as highly reliable as the wings and fuselage of modern aircraft. Protection against radiation, without incurring an excessive weight penalty, also enters the picture.

Speaking as a biologist, H. Strughold (*op. cit.*, Note 29) notes the existence of a physiological dividing line between atmosphere and space that is well within the limits of the Kármán line. He sees as decisive from the biological point of view the fact that we face complete anoxia at about 52 000 feet (16 km) despite the occurrence of free molecular oxygen up to the region of the Kármán line.

R. M. Salter (*op. cit.*, Note 29) states that the air-breathing vehicle is limited in altitude, citing as an example that at 20 miles (32 km) the required Mach number for a ramjet is more than 5 and the result incoming air has a stagnation temperature of the order of 2000°F. Since energy must be imparted to this air at higher temperatures, Salter says, it may be seen that a present engineering limitation on suitable fuels and materials is approached.

The basic advantage of a criterion such as the Kármán line lies in its practical application — it effectively separates the territory of air-breathing vehicles from that of rocket vehicles.

The same cannot be said for limits set in chemical terms of where airspace or atmosphere ends. Thus, Kucherov (laying aside scientific discipline) and citing Joseph Kaplan's statement that "at 250 miles there is less air than in the best vacuum tube obtained on earth", concludes that "the exosphere 250 miles above the earth is no longer 'air space' . . . and must be deemed free of sovereignty projected from the earth" [13].

Writing in the earlier colloquium, but dealing with this type of solution, John Cobb Cooper understandably contends on the basis of definitions confused by scientists that "upper 'air space' boundary will be above much satellite and most guided missile flight, and will be a most uncertain line" [14]. Indeed, the top of the "atmosphere" has been estimated at anywhere from 10 to 650 miles above the earth's surface, depending upon the obfuscated viewpoint of the "scientist" discoursing.

Numerous authentications of the validity of the Kármán line may be found in the legal and scientific literature of every nation in the world. Just two examples suffice. In the USSR the Latest University Textbook on International Law by Lisovskiim V. I. *Mezhdunarodnoe pravo,* Kiev, 1955, p. 159, states:

> "The sovereignty of a State extends to the stratosphere (that is, the stratum of air from 11 to 75 kilometers from sea level) which is over the territory of the State as well as to the troposphere (that is, stratum of air up to 11 kilometers from the sea level) over this territory because at the present level of technique of aviation not only peaceful flights may be performed in the stratosphere, but also military operations. Consequently, the State must have the right to regulate the traffic of the foreign vessels also in this stratum of the air."

Representatives of the USA and the USSR in connection with arriving at agreements within the Federation Astronautique Internationale as to records in both air flight and space flight have agreed upon the definition of space flight as being flight above 62 miles (100 km) altitude. At this point aircraft flight must end and space flight begin. This agreement coincides with the Kármán line theory.

Actually, we are talking about sovereignty and jurisdiction — *not* the nature of vehicles. We cannot fly any aircraft, *without permission,* over the airspace of another sovereign nation, and we cannot launch a Mars-bound rocket vehicle through the airspace of another sovereign nation, nor can any other nation launch a space vehicle through our airspace. These propositions are universally observed — even through our launching ranges.

Cooper's survey of proposed solutions gives qualified approval to the Kármán line, noting that it "is capable of physical and mathematical demonstration at a reasonably stable height", but citing criticism that the line will vary with change in climatic conditions. This criticism is effectively answered by referring to the hundreds of millions of radio measurements which have established a master curve extablishing a median interference factor.

Cooper himself (44 *Am. Bar Ass'n. Journal* 321, April 1958) has declared that the airspace over which states have sovereignty "includes only areas where sufficient gaseous atmosphere exists to provide aerodynamic lift for such flight instrumentalities as balloons and aircraft". He refers to outer space as "beyond the territorial sphere of any state", but cites the lack of agreement on the boundary between territorial airspace and outer space, or on the legal status of the "intermediate area" in which "the presence of a certain amount of gaseous atmosphere may cause the fall of flight instrumentalities, thus endangering the state below".

Still earlier, Cooper proposed an international convention which would, *inter alia,* "extend the sovereignty of the subjacent state upward to 300 miles above the earth's surface, designating this . . . area as 'contiguous space', and provide for a right of transit through this zone for all nonmilitary flight instrumentalities when ascending or descending" [15]. In a letter to the London *Times* on September 2, 1957, Cooper extended the limit of his "contiguous space" to 600 miles. This, of course, is purely a political, and not a legal or scientific concept.

Dr. Eugène Pépin and C. Wilfred Jenks, on the other hand, would limit national sovereignty to its present boundaries in the atmosphere; although they do not define these in quantitative terms, their theses are compatible with the Kármán line.

Pépin says "it should be taken for granted that over and around the surface of the earth (land or sea areas) there is what the scientists call 'atmosphere', over certain parts of which national sovereignty is extended; and above the atmosphere there is 'space' . . . Therefore, from a legal point of view, there are only two zones: one, the air or atmosphere, which has a legal status already defined in an international instrument, and the other, the space of a still undefined status" [16].

Jenks states: "Space beyond the atmosphere is a *res extra commercium* incapable by its nature of appropriation on behalf of any particular sovereignty" [17].

Jenks proceeds to urge that "jurisdiction over activities in space beyond the atmosphere should be recognized to be vested in the United Nations and that legislative authority over activities beyond the atmosphere of the earth should be exercised by the General Assembly acting through or on the advice of an appropriately constituted body".

C. E. S. Horsford (*Journal of the British Interplanetary Society*, May-June 1955, 144—150) considers the sovereignty concepts of the Chicago Convention "inadequate" and "largely inapplicable" to the medium of space. "All operations in space", he points out, "will be conducted so far above what is now accepted as the airspace above a nation's territory, and so impossible will it be to observe any limitations of a territorial nature such as frontiers demand, that it is in the law of the sea that the answer would seem to lie."

Turning to maritime analogies, Horsford reasons that "in the light of modern international theory, outer space itself is likely to be considered a free navigable area as are the high seas . . ."

Alex Meyer lines up with these spokesmen against extension of sovereignty concepts to outer space. Taking issue with both Becker of the USA and Galina of the USSR — *infra*, [20 and 21] — this air law specialist from the German Federal Republic highlights the impossibility of establishing in outer space a defined territory which would correspond to the boundaries of a particular state on earth. Even if the laws of nature could be amended to achieve this dubious end, Meyer joins Professor Lissitzyn in the belief that it is also impossible to establish effective control over an area thus defined in outer space. The rights of sovereignty, he concludes, cannot be realized in outer space. He offers solace to those concerned about national security, however, by granting that states may engage in outer space activities on the basis of rights other than sovereignty, *e.g.*, the right of self-defense.

Bin Cheng concludes his inquiry into the legal status of flight space (*Int. and Comp. L.Q.*, July 1957) by observing that "in view of the fact that assertions of national sovereignty at extreme heights become increasingly less definable and meaningful even for states of continental dimensions, all that may be said is that a not unreasonable conjecture is that states will subject the whole of outer space to the same regime as the high seas, with possibly an extended contiguous zone consisting of the outer fringes (the exosphere) of the terrestrial atmosphere".

Cheng's carefully phrased statement may be construed as placing him in the above company of those who oppose extension of national sovereignty to outer space, but would allow its free use above a certain line, perhaps the Kármán line.

Dr. Welf Heinrich Prince of Hanover expresses the view that of the many attempts made to define the principles by which terrestrial sovereignty should be restricted in a vertical direction, "the solution recommended by Mr. Haley and Dr. von Kármán is best suited to serve as the basis of an international convention on the subject" [18].

The importance of reaching some common ground is underlined by Professor Cooper: "To me, it is a most practical and urgent question that there be early international agreement defining these areas. The problem presents more dangerous possibilities than the fixing of the boundary between national marginal seas and the high seas. Yet states in the international community have deemed it urgent to agree to those boundaries. If discussion of the limits of airspace and outer space is too long delayed, the problem will rapidly become almost insoluble" [19].

Nevertheless, the positions expressed by official or quasi-official spokesmen of the powers which lead in space exploration have been on the cautious side.

Loftus Becker, the Legal Adviser of the US State Department, wrote as follows: "Although the US has plainly asserted its complete and exclusive sovereignty over the airspace above its territory, we have at no time conceded that we have no rights in the higher regions of space" [20]. Here he is speaking as a statesman and diplomat, invoking the political article of the United Nations covenant, but, of course, not as a lawyer.

Similarly, a Soviet spokesman has stated that "the outer altitude of space sovereignty must be established in such a manner as to protect the state against encroachments on its territorial sovereignty" [21].

It is encouraging to note, however, that the several satellites launched by the USSR and the US during and following the International Geophysical Year orbited the earth above the Kármán primary jurisdictional line without evoking claims that "national space" had been violated. This writer first advanced this legal theory long before Sputnik I [21 a].

It is encouraging also that the great powers, while remaining cautious about yielding any rights in space which they may have, are not asserting definitive claims which would freeze them in cold war postures of opposition and preclude any future agreement. (A similar forbearance has led to the signing of an Antarctic treaty, now before the legislatures of the signatory nations for ratification.) Instead, national views appear to be going through a process of development and maturation. True, the development is not uninfluenced by changing rates of progress in the space art, but maturation results as it begins to appear that no nation can achieve dominance in the vast realm of space.

Thus in 1956, following the launching by the US of high-altitude weather balloons, we find A. Kislov and S. Krylov citing with approval the 1913 dictum of Clunet that "sovereignty of each country over its territorial air space must . . . extend *usque ad coelum*", *i.e.*, they added, "endlessly" [22].

But on October 17, 1957, following the success of Sputnik I, G. P. Zadorozhnyi, Doctor of Juridical Science, wrote, "By analogy to the principle of freedom of the open seas, which beyond the limits of territorial waters and special maritime zones do not belong to anyone and are in general use by all nations, the upper atmosphere, which is beyond the limits of effective air control by states, can likewise be considered a zone of open air, in general use by all nations" [23].

A similar, but slightly qualified advocacy of freedom in outer space, is offered by A. Galina in 1958. "The upward limit of state sovereignty must be established in such a way that the state is protected against attempts to violate its territorial sovereignty and its independence . . . but to recognize that no state has the right to subject parts of cosmic space to its legislation, administration, or jurisdiction" [24].

Also in 1958, F. N. Kovalev and I. I. Cheprov assert that state sovereignty can be extended into the legal vacuum they admit exists in outer space, but only as far as the state can effectively enforce it. Moreover, they caution states against making exorbitant claims that would endanger scientific exploration.

Soviet thinking is perhaps most comprehensively summed up by E. Korovin ("International Status of Cosmic Space", in *International Affairs,* 53—59, January 1959). After rejecting attempt to draw analogies from the Chicago Convention or from maritime and air law, Korovin concludes:

"Most scholars tend for a variety of reasons to deduce that national sovereignty cannot be extended to the Cosmos (outer space) and thereby reject the right of any country to put cosmic space under its legislation, administration or jurisdiction. This applies to such well-known experts in the capitalist countries as Haley and Cooper (USA), Jenks and Horsford (Britain), Meyer (Federal Republic of Germany), and such jurists from the Socialist countries as Makovski (Polish People's Republic), Gabrovsky (Bulgaria), Reintanz (the German Democratic Republic), Soviet authors (Galina, Zadorozhni, Kovalev).

"There are also good astronomical reasons to back this viewpoint. In view of the Earth's rotation and velocity of the Earth's motion and the whole solar system, any one point of cosmic space may be 'over' a particular country (in 'its' cosmic cone) only for an insignificantly short time, frequently just a fraction of a second. Sovereignty in the Cosmos may thus be enforced only at lightning speed and in continuous movement . . .

"It must be concluded that national sovereignty does not extend to cosmic space. However, it does not follow that space is to be considered as some kind of legal 'vacuum' where no restrictions on freedom of action prevail. All universally accepted rules of international law (inadmissibility of the use of force in solving international disputes, noninjury of foreign citizens and their property, governmental responsibility for the activities of their representatives, *etc.*) apply to the Cosmos as well."

What is the viewpoint of one who approaches the question from a supranational background? Oscar Schachter, Deputy Director of the United Nations Legal Department, took a position in 1952 which in general may be bracketed with those of Cheng, Haley, Hanover, Jenks, Meyer, and Pépin. Declaring that in international law the term "airspace" is found in aviation treaties and therefore is "presumably intended to refer to the part of the atmosphere which contains enough air to allow aircraft to fly", he calls for fixing the limit of national sovereignty at the upper boundary of the airspace. Beyond the airspace, "outer space and the celestial bodies would be the common property of all mankind, over which no nation would be permitted to exercise domination. A legal order would be developed on the principle of free and equal use, with the object of furthering scientific research and investigation" [25].

This is, of course, an individual opinion. The United Nations officially has taken no position on a legal distinction between airspace and outer space. A report made in June 1959 by the UN *Ad Hoc* Committee on Peaceful Uses of Outer Space deems any attempts at such official definitions premature at this time. It did suggest, however, that one approach to the problem would be to establish the limits of airspace and outer space within a practicable range, and that the type of space activity could be explored as a basis for legal control [26].

George J. Feldman, a participant in the work of the Committee, takes issue with its omission of the delimitation problem from the list of priority problems. He comments: "It may be true, as the Committee says, that an international agreement based on current knowledge and experience would be premature. Again, however, the important thing is to begin now — conduct research, make studies and investigations, and work out the principles on which an agreement may ultimately be based (subject to change in the light of later knowledge)." Feldman also suggests that by changes it made in a draft report, the Committee gave implicit approval to treating outer space as *res communis omnium* [27].

The Committee was explicit enough in expressing its belief that with the unchallenged launching of space vehicles during and after the IGY, "there may have been initiated the recognition or establishment of a generally accepted rule to the effect that, in principle, outer space is, on conditions of equality, freely available for exploration and use by all in accordance with existing or future international law or agreements".

And the report (United Nations General Assembly Document A/AC.98/2) did say that "although an attempt at comprehensive codification of space law was thought to be premature, the Committee also recognized the need both to take timely, constructive action and to make the law of space responsive to the facts of space".

On the matter of whether study of the legal problems of space is premature, Pépin offers an interesting historical sidelight (in a lecture, April 20, 1959, at Institute of Air and Space Law, McGill University, Montreal). He reminds us that in 1900, three years before the first flight of the Wright brothers, the Institute of International Law discussed a draft convention on the legal status of aircraft. Like Jenks, Pépin concludes that "if we want to reach agreement on questions affecting all states in the world, we cannot wait until situations become crystallized, in fact, nor until rules of custom — which are often difficult to modify — have been established".

It is also interesting to note the treatment of this subject in the Staff Report of the Select Committee on Astronautics and Space Exploration, established by the 85th Congress of the United States with Feldman as Director and Chief Counsel.

The report first considers the position of experts in the United States and elsewhere who champion a case-by-case approach. For example, Rear Admiral Chester Ward, Judge Advocate General of the US Navy, declares "we are being distinctly premature if we attempt to set up or to propose specific rules of space law at this early stage" [28]. (A more detailed exposition of this point of view is given by McDougal and Lipson, *Am. Journ. Int. Law*, 407—431, July 1958. The 1958 report of the Committee on the Law of Outer Space of the Section of International and Comparative Law, American Bar Association, also falls in the "wait and see" category.)

The Staff Report then examines the opposing view of those who hold with Jenks that "the possibility of developing the law on sound principles depends primarily on an initiative being taken in the matter before *de facto* situations have crystallized too far". Weighing the contrasting stands, the report finds eight reasons why the latter position is more convincing [29].

Stephen Gorove (in *New York Law Forum*, 305—328, July 1958) takes cognizance of wide agreement that "elongation of national sovereignty into the limitless spheres of the universe is ... untenable" and that it "should not reach beyond the airspace". He sees the future development of space law, however, as hinging less on resolution of the disputed extent of the airspace than on the decision-making factors of world power politics. To Gorove, therefore, an acceptable international inspection system is the most urgent problem.

A clue to the direction of development for international space law may be found in the development of domestic air law. The very doctrine of national sovereignty over the airspace which a handful of commentators would expand to all outer space is the relatively recent expression of the rights of a larger group, the nation, over the once supreme individual property owner. The necessities of modern air transportation have caused the courts to declare those superior rights. Thus the Supreme Court of the United States in *US v. Causby,* a 1946 case [30] which is a landmark in US air law, held: "It is ancient doctrine that, at common law, ownership of the land extended to the periphery of the universe — *cuius est solum eius est usque ad coelum.* But that doctrine has no place in the modern world." May it not eventuate that a still larger group, the family of nations, will one day declare its dominance in the realm of space law?

We have touched in this paper on only one phase, though a crucial one, of the establishment of the rule of law in outer space. In addition to fixing the terminal point of terrestrial jurisdiction, we must cope with many other legal problems peculiar to the space age. For example, if we succeed in delimiting outer space and seek to confine its use to peaceful purposes, how do we define "peaceful"? Does it mean "non-military", or as Feldman suggests, "non-aggressive"?

If a space vehicle causes injury, what kind of tort law should be applied [32]?

How shall we accommodate the use of transmission frequencies in space to the rules and requirements of ground radio services [33]?

Ownership and use of the moon may already have moved from the category of a remote question to an imminent one as a result of the successful launching in September 1959 of Lunik II [34].

These and a host of other questions [35] are beyond the scope of the present treatment. Their diversity and complexity, however, suggest the importance, indeed the urgency, of accelerated progress toward answers which are both sound and generally acceptable. Space science and technology move forward at hypersonic speed. The law cannot afford to remain earthbound. The mildest possible penalty for such a lag will be confusion. The maximum price we may pay is mutual destruction.

What are some of the immediate legal problems?

1. Within the framework of the pertinent international treaties, lawful use must be made of radio frequencies for all forms of astronautical communications. This requirement of international law has been observed only once by any nation since the launching of Sputnik I. I venture so say that as time goes on the international obligations of the nations of the world will be more and more ignored — if the lawyer does not intervene and make himself heard in the United Nations and in the International Telecommunication Union. Many radio frequencies are needed for communication between earth and vehicle in space, and between vehicles in space and earth; between earth and positions in space, and positions in space and earth; between two or more positions in space; between two or more space vehicles. Radio frequencies are essential, not only for all forms of communication between the fixed and mobile points I have thus indicated, but also for numerous other purposes such as for telemetering, tracking, guidance, radiopositioning (radar), and so on.

2. Any nation sending radio equipment into space (except equipment destined for probes beyond Mars and Venus) must be required to be able to command such radio equipment to stop transmitting — or the equipment may be the source of interfering signals for decades to come. With the improvements in solar batteries

and the use of outer orbits where sunlight is constantly available as the power source, radio equipment in satellites may well, in a very short time, be capable of indefinite life, and therefore of indefinite interference, unless controllable.

3. No object should be placed in any orbit in outer space which cannot be guided back to earth or destroyed by some other means, such as being guided into the surface of the sun. The nations of the world — including the USSR, USA, United Kingdom, Peoples' Republic of China, and so on, contemplate sending scores of "Sputniks" into space. Many of these undoubtedly will attain permanent orbits. Remember, it takes about as much energy to get one of these objects back to earth as it does to place it into orbit initially. As a practical matter, it would be almost impossible to divert an earth-orbiting object while outside the earth's atmosphere, without having placed on the object initially a mechanism with which to divert the object. One cannot destroy the object by ordnance, as even if blasted the fragments would continue on orbiting and probably become even a greater menace to navigation and safety in space. Therefore, we must now enact international regulations requiring that before any object is sent into space it must be equipped with apparatus whereby it may be commanded back to earth at a safe location on earth.

4. By the same token, any object sent into space must be under the control of the sender so that on completing its orbital life the responsible party may guide the object back to an area safe for mankind. In other words, as "Sputniks" become larger there is no assurance whatsoever that they will atomize on their return to earth. The fact is that many of these objects will come back in large and lethal metallic chunks and there is always the possibility that these metallic monsters will hit congested population areas. Assumptions of adverse odds really afford no criteria of safety. Only a few years ago, in America, and specifically at Nellis Air Force Base in Nevada, it was customary for jet planes to describe a run over a certain radio path, ascend high into the sky, and then perform the penetration dive. It was believed that this maneuver was entirely safe, that the odds that any injury would occur to commercial aviation were so remote that notice should not be taken thereof. However, a United Airlines DC-7 in full and cloudless daylight, engaged in a routine scheduled flight, with more than 40 passengers aboard, was crashed into by one of these jets during the course of the penetration dive, and all aboard each craft were killed. We cannot assume that the returning debris of numerous "Sputniks" will not cause damage on earth. It is fundamental that safety precautions be enacted and enforced. Of great importance to future manned space navigation is keeping the "space ways" clear for safety of life and property in space. This means we must provide now against all forms of space derelicts.

5. At the present time intercontinental ballistic missiles and missiles of lesser range are being tested over wide areas of the earth's surface, and by several nations. With respect to all such firings it should be required that flight plans be filed with an appropriate agency such as ICAO, or a new office in the United Nations, for the advance knowledge and guidance of surface craft and installations and for the general "peace of mind" of humanity.

6. All long-range missiles and all "Sputniks" should be required to carry apparatus which will render them readily identifiable.

7. Agreement among nations with respect to the use of television, photography, and any observational equipment whatsoever, should be immediately undertaken.

8. There should be organized within the United Nations or by universal agreement as a result of special treaty making, a Commission to define the limitations of sovereignty of the nations of the earth. Two problems are involved: (a) stating the present international law with respect to limitations on jurisdiction, and (b) reconciling final determinations with the statutes of the individual nations of the

earth. (This means that the individual statutes of each of the nations of the world relating to jurisdiction and national sovereignty may have to be changed.) I have pointed out in numerous lectures and articles that all existing treaty law and all existing national statutes define the limit of national jurisdiction to be within "air space". There simply is no law for the area beyond "air space", which we now call "outer space". There seems to be, on the part of the Foreign Ministries of many nations, a strange indifference to this question of defining jurisdiction. But there is no way to escape the problem. I will not labor the point on this occasion, but I would like to leave with you the thought that this is one of the most urgent and important of all problems. We are occasionally faced with strange and tormented arguments such as those concerned with the nature of the X-15. Some persons claim that because it is airborne for a portion of its flight, such fact has some sort of continuing jurisdiction implication. This is not the case. As long as it is airborne it is subject to the jurisdiction of the nation in the air space over which it travels. When it is no longer airborne — and is beyond air space — and is traveling by centrifugal force, the object is in outer space and no present law appertains to its movement. The same precepts are true with respect to any objects (such as Sputniks) which are hurtled into outer space. This is a situation which must be handled legally in the near future.

9. My colleague, Dr. Welf Heinrich Prince of Hanover, and I talked to the students and faculties of 26 great American Universities concerning the problems of jurisdiction and sovereignty in positions in outer space such as the moon, Venus, Mars, and so on; and also with respect to mobile vehicles and orbiting objects in outer space. We pointed out that through the medium of the United Nations or by universal treaty making, at this time the moon and other natural objects should be placed beyond the jurisdictional or sovereignty claims of any nation on earth. The moon may be alighted upon by human beings within the next five or ten years, and if in the meantime we do not reach an understanding concerning the moon the nation achieving this great scientific acquisition may well, under classical principles of terrestrial international law, claim sovereignty over the moon. Great military leaders have said that the nation which controls the moon will also control the earth. The policy of inaction and inattention to this problem may well haunt those who are responsible for such inaction. If Nation X establishes a base on the moon, to the dismay of Nation Y, it will be found that the numerous small nations of the earth will really sympathize with Nation X because of its great achievement and on the grounds of fair play — and will afford little sympathy to Nation Y, which has lagged behind, which has not taken the initiative in guarding against the very problems which would arise, and which now cries "mother".

10. In the background of all the foregoing, as I have pointed out since my American Rocket Society talk in 1954, we must face the space law problems of drafting, administering and enforcing regulations relating to safety, sanitation, health, asylum, equipment, navigation, emigration and immigration, all of which regulations would conform to the most universal and enlightened principles of freedom and the use of property, and promulgate a code defining public and private liability for damage.

References

[1] Reprinted in Space Law — A Symposium, Prepared at the Request of Honorable Lyndon B. Johnson, Chairman, Special Committee on Space and Astronautics, United States Senate, Eighty-Fifth Congress, Second Session, December 31, 1958, by special permission from Foreign Affairs, October 1958, pp. 95—106. (Copyright by Council on Foreign Relations, New York. The author, *Phillip W. Quigg*, is Assistant Editor of Foreign Affairs.)

[2] See for example the American Bar Association Report to the National Aeronautics and Space Administration on the Law of Outer Space, reported by *Leon Lipson* and *Nicholas de B. Katzenbach,* Chicago, August, 1960.

[3] *Andrew G. Haley,* Rocketry and Space Exploration, pp. 126—27. (D. Van Nostrand Company, Inc., 1958.)

[4] See description hereinafter and also in *Appendix 5.*

[5] NASA Release, June 16, 1959. Address by *Robert E. Horner,* Associate Administrator, NASA, to the Second Annual Industry Missile and Space Conference, Detroit, Michigan.

The statement of *Dr. Hugh L. Dryden,* Deputy Administrator, NASA, delivered before the Western Space Age Conference at Los Angeles, Calif., NASA Release, March 5, 1959. This offers a most interesting contrast to the Horner statement, in that the differences between "air plane" and "space vehicle" are explained with lucidity. Dr. Dryden described the X-15 as follows:

> "We now have decided also to proceed from the missile by direct method of catapulting a manned capsule into orbit, above the aerodynamic limits of the atmosphere, and bringing it down with a suitable recovery system. The first approach is represented by the X-15, the new experimental rocket craft, which was built in California and is now undergoing its preliminary trials at Edwards Air Force Base, in the California desert."

[6] Only too many scientists seem to be embroiled in "hang over" terminology. Thus, many use the terms "air" and "atmosphere" as meaning anything from cosmic plasma to hydrogen clouds in distant nebulae. "Air" and "atmosphere" are terms of art and the meaning of these words is well known (see *Appendices* 1, 2, and 3). The scientists must, for the sake of accuracy expand their technical vocabulary to make their statements understandable and precise.

[7] NASA Release, May 19, 1960, Statement by Dr. H. L. Dryden.

[8] *Appendix* 6 attached hereto.

[9] *Andrew G. Haley,* "Recent Developments in Space Law and Metalaw", Harvard Law Record, February 7, 1960.

[10] See *M. Smirnoff,* "The Need for a New System of Norms for Space Law and the Danger of Conflict with the Terms of the Chicago Convention", in First Colloquium on the Law of Outer Space, Haley and Hanover, eds., (Springer-Verlag, Vienna 1959). See *Appendices* 1, 2 and 3.

[11] See *H. Strughold* and *E. O. Hulburt* contributions to "Physics and Medicine of the Upper Atmosphere" (University of New Mexico Press, Albuquerque, 1952). A more recent work — Massey and Boyd, "The Upper Atmosphere" (Philosophical Library, New York, 1959) — points out that "the actual proportion of oxygen present in the atomic form at any height is difficult to calculate because ... of mixing forces due to atmosphere motions and to diffusion. These prevent the proportion of diatomic (*i.e.,* molecular) oxygen from falling off at great heights as rapidly as would be expected from their formulae, under which the atomic concentration at 66 miles would be nearly 100 times the diatomic."

[12] *Haley,* "Space Age Presents Immediate Legal Problems", in First Colloquium — *supra,* [10].

[13] *Samuel Kucherov,* "Legal Problems of Outer Space", Second Colloquium on the Law of Outer Space (Springer-Verlag, Vienna 1960).

[14] *Cooper,* "The Problem of a Definition of 'Air Space' ", in First Colloquium — *supra,* [10].

[15] *Cooper,* "Legal Problems of Upper Space", 23 J. Air L. 308 (1956).

[16] *Pépin,* "Space Penetration", speech at Annual Meeting, American Society of International Law, April 1958.

[17] *Jenks,* "International Law and Activities in Space", 5 Int. and Comp. L. Q. 99 (1956).

[18] *Hanover,* "Circle of Thoughts", Second Colloquium, *supra,* [13].

[19] *Cooper,* at American Bar Association, Miami, August 1959.

[20] *Becker,* "Major Aspects of the Problem of Outer Space", Department of State Bulletin, June 9, 1958.

[21] *A. Galina,* "On the Question of Interplanetary Law", Soviet State and Law, 7, 52—58, (1958).

[21a] 23 Harvard Law Record, Special Supplement pp 1—2, November 8, 1956.

[22] *Kislov* and *Krylov,* "State Sovereignty over Air Space is an Acknowledged Principle of International Law", International Affairs, *3,* 34—43 (1956).

[23] *Zadorozhnyi,* "The Artificial Satellite and International Law", Sovietskaia Rossiia *246, 3.*

[24] *Supra,* [12].

[25] *Schachter,* "Who Owns the Universe?", from the book, "Across the Space Frontier". (C. Ryan, ed., Viking Press, New York, 1952.)

[26] *E. Galloway,* "The United Nations *Ad Hoc* Committee on the Peaceful Uses of Outer Space", Second Colloquium, *supra,* [4].

[27] *Feldman,* "The Report of the United Nations Legal Committee on the Peaceful Uses of Outer Space: A Provisional Appraisal", *ibid.*

[28] *Ward,* "Projecting the Law of the Sea into the Law of Space", JAG Journal, March 1957, 3—8.

[29] Survey of Space Law, Staff Report of the Select Committee on Astronautics and Space Exploration (U.S. Government Printing Office, Washington, 1958).

[30] 328 U.S. 256, 260—261.

[31] *Supra,* [18].

[32] *de Rode-Verschoor,* "The Responsibility of States for the Damage Caused by Launched Space-Bodie", *supra,* [1], *inter alia.*

[33] *Haley,* "Space Communications — A Decade of Progress", American Rocket Society, Los Angeles, May 9—12, 1960.

[34] *Cocca,* "Principles for a Declaration with Reference to the Legal Nature of the Moon", *supra,* [10].

[35] See *Hogan,* A Guide to the Study of Space Law, 5 St. Louis U. Law Journal 79 (1958).

Appendix 1

AIR, *noun. 1.* The gaseous substance surrounding the earth, being principally a mixture of gases (although often considered a single gas), and consisting mainly of nitrogen and oxygen, in the ratio of about four parts of the former to one of the latter, but containing also varying amounts of water vapor, and relatively small quantities of the gases argon, carbon dioxide, hydrogen, neon, helium, krypton, and xenon. Particles of dust and smoke, bacteria, spores, etc. suspended in the air are not usually considered a part of it. Air extends upward with decreasing density from the surface of the earth, having a normal pressure of about 14.7 pounds per square inch at sea level. Air is both compressible and elastic, and its principal importance in aeronautics is its character as a fluid, thus affording a medium or means of support for aircraft. *2.* Any particular portion of this substance, distinguished by some special quality, character, or treatment from the surrounding air, as *air* that migrates from the polar regions, or, a leakage of *air* from a cylinder. *3.* A gas — in this sense esp. in some combinations and attributive uses; *4. a.* The air (in sense *1*) regarded as the realm or medium in which aircraft travel or operate. *b.* This air conceived of as a medium for commerce, warfare, etc. by the use of aircraft — disting. from the land and sea, as, transport by *air,* or, the conquest of the *air. 5.* A military force or service that operates in the air (sense *1*), or military force or power exterted in or from the air (air power), as, the defeat of the enemy *air,* or, the coordination of land, sea, and *air.*

Adams, Frank Davis. *Aeronautical Dictionary,* 1959, p. 7 (Government Printing Office, Washington, D.C.).

Appendix 2

ATMOSPHERE. The term "atmosphere" usually refers to the gaseous envelope covering the surface of the earth. The word is derived from the Greek words ἀτμός (smoke or vapour) and σφατρα (globe or sphere). The early Greeks were probably the first to study the weather in a regular and systematic way and the wind was defined by Anaximander as a "flowing of the air". Hesiod in his treatise "Works and Days" discussed the origin of wind, and many observations of physical properties of the air were made by Ctesibus,

Hero of Alexandria, and others. The material nature of air is clearly recognized in Hero's "Pneumatica".

Anaximenes (c. 500 B.C.) regarded the air as the primordial substance from which all matter was condensed. During the time of Socrates meteorology was neglected, but Aristotle revived interest in the study of the atmosphere and wrote about the winds. He regarded the atmosphere as consisting of three regions; the lowest in which plants and animals exist he supposed to be immovable like the earth; the uppermost region adjoined the fiery heavens and moved with them; the division intermediate between the other two, he believed to be exceedingly cold. Meteors were considered by Aristotle to be exhalations from the earth, which became incandescent when they reached the hot upper layer.

Very little progress was made from this time until the early part of the 17th century, although it is said that during the 11th century the Arabs calculated the height of the atmosphere, from the duration of twilight, as 92 kilometers. In 1643, Torricelli, a student of Galileo, found that if a long glass tube sealed at one end was filled with mercury and the open end closed with the finger while the tube was inverted in a vessel containing mercury, the liquid sank only to a certain level. It thus became possible to measure the pressure of the atmosphere, and the space above the mercury is still referred to as a Torricellian vacuum. This apparatus was called a barometer (q.v.) by Boyle and soon came into general use. Pascal demonstrated the decrease of the pressure of the air with altitude by measuring the height of the mercury column of a barometer at different points up a tower in Paris. In 1650 von Guericke (q.v.) found that he could pump air and was responsible for the famous experiment with the Magdeburg hemispheres.

That air consists chiefly of two gases was first recognized by Scheele (1772), but Cavendish (1781) was responsible for a larger number of analyses of the air and found that 100 volumes contain 20-83 parts by volume of oxygen and 79-17 of nitrogen. Similar experiments were carried out by Priestly (who thought the composition variable) and Lavoisier, but it was not until 1846 that it was definitely established by Bunsen that the composition of the atomosphere is not absolutely constant.

The Composition of the Atmosphere. — Air is a mixture of gases and is not a chemical compound. This is proved by the following: — (*1*) The composition of air is not constant, and the quantities present of the different components do not bear any simple relation to their atomic weights. (*2*) The constituents can be separated by diffusion and by the fractional distillation of liquid air. (*3*) Air dissolves in water in accordance with the law of partial pressures and hence air expelled from water contains an increased proportion of oxygen.

Below a height of 20 km (12½ mi) the constituents of the atmosphere, with the exception of water vapor, are well mixed by winds and by diffusion. Slight changes in composition do occur, however, at the surface of the earth and these depend on latitude and the presence of large quantities of vegetation or sea-water. The permanent constituents of the air are generally present in the following proportions (according to Humphreys in the "Scientific Monthly", 1927):

Substance Total atmosphere	Volume % in dry air
Dry air	100.00
Nitrogen	78.03
Oxygen	20.99
Argon	0.9323
Water vapor	
Carbon dioxide	0.03
Hydrogen	0.01
Neon	0.0018
Krypton	0.0001
Helium	0.0005
Ozone	0.00006
Xenon	0.000009

The following table by Hann shows the variation with latitude.

	Nitrogen	Oxygen	Argon	Water vapor	Carbon dioxide
Equator	75.99	20.44	0.92	2.63	0.02
Latitude 50 N.	77.32	20.80	0.94	0.92	0.02
Latitude 70 N.	77.87	20.94	0.94	0.22	0.03

The composition also varies altitude, but not to any very appreciable extent at heights at which respiration is still possible. The amount of water vapor present in the air is usually about 1.2 % by volume, but in very cold weather this quantity falls almost to zero. At other times it may be as high as 5 %.

Height of the Atmosphere. — The height to which the atmosphere extends cannot be definitely stated, although at an altitude of 50 mi the air cannot exert any measurable pressure. Three methods are available for the estimation of the height: (1) observation of meteors, (2) measurement of the duration of twilight, (3) observation of auroral displays. The first method gives results ranging from 150 to 300 km, while the duration of twilight indicates a value of about 64 km at lat 45°. It is difficult to make reliable calculations from auroral displays, but it is claimed that these occur up to a height of 500 km. If density of the atmosphere remained uniform throughout with the same value as at the earth's surface, the air would form a layer only 8 km thick and this is sometimes called the "height of the homogeneous atmosphere". Half of the air is below a height of 5—8 km. At low levels temperature is usually considered to decrease 0.56°C per 100 m increase in altitude, but the rate is extremely variable. Above 2 km the temperature is on an average below 0°C and continues to fall up to 10 km (6 mi) when it is about —55°C. At 37 km the temperature is practically the same as at 10 km. The lower region of the atmosphere is known as the "troposphere" and extends up to 10 km, beyond which clouds are not generally found, except in tropical latitudes.

Absorption of Radiation by the Atmosphere. — The blue color of the sky is due to the fact that the air is not perfectly transparent and its particles reflect and scatter light, that from the blue end of the spectrum being most widely scattered. This effect also obscures the light of the stars. Very little of the sun's thermal radiation is absorbed by the air, which derives most of its heat from the earth by conduction and convection. A layer of air one metre thick absorbs about 0.007 % of the radiant heat passing through it. Of the radiation incident on the outer atmosphere about 37 % is lost by reflection and scattering. The fraction of the radiant energy from the sun which reaches the earth is termed the coefficient of transparency of the atmosphere. The absorption is chiefly dependent on the amount of water vapor, carbon dioxide and solid impurities present and consequently is much greater in the neighborhood of towns. The following coefficients of transparency are given by Wild for one meter of air: Dry, dust-free air, 0.99718; dry air containing dust, from a room, 0.99520; dust-free air saturated with water vapor 0.99328. The ozone, which appears to be present at very high altitudes, is responsible for the removal of practically all the ultraviolet radiation of wave-length shorter than $\lambda = 2885$ A.U.

Since the temperature of the upper atmosphere is practically constant and no convection or condensation takes place there, it is important to consider what would be the effect of dust particles which might be forced into the stratosphere by volcanic eruption. After certain eruptions, *e.g.*, Krakatoa 1883, Mont Pele and Santa Maria 1902, Katmai 1912, a reddish halo was observed round the sun owing to the dust ejected to very great altitudes, and it was possible to calculate the size of the particles. It has been estimated that a quantity of dust of volume less than $1/174$ km^3 distributed in the upper layers of the air, would reduce the intensity of solar radiation by 20 %. It is possible to explain the occurrence of ice ages in this way.

Encyclopaedia Britannica, Vol. 2, 1943.

Appendix 3

AIR (*1*) The mixture of gases in the atmosphere. (*2*) The element that gives lift to air-craft, or offers resistance to objects that move through it. (*3a*) The region above and around the earth, including the atmosphere and the space beyond, subject to control by air or space vehicles, in contradistinction to land and sea. (*3b*) That part of this region that includes the atmosphere up to its effective upper limits, but not outer space.

ATMOSPHERE. The body of air which surrounds the earth (or any other celestial body), defined at its outer limits by the actual presence of air particles but in such few numbers that collisions between them are so rare as to make the force of gravity the only means of keeping them associated with air particles at lower altitude.

> Gaynor, Frank. *Aerospace Dictionary*, pp. 10 and 25 respectively. (Philosophical Library. New York, 1960.)

ATMOSPHERE, *Composition of* (Chem, *etc.*). Dry atmospheric air contains the following gases in the proportions (by weight) indicated: nitrogen, 75.5; oxygen, 23.2; argon, 1.3; carbon dioxide, 0.05-0.4; krypton, 0.029; xenon, 0.005; neon, 0.00086; helium, 0.000056.

> *Chambers's Technical Dictionary*, p. 57. (Ed. By C. F. Tweny and L. E. C. Hughes. W. & R. Chambers, Ltd. London, 1954.)

Appendix 4

BIBLIOGRAPHY OF BIBLIOGRAPHIES ON SPACE LAW

1. Association of the Bar of the City of New York, Checklist of Materials on Law and Outer Space, The Record of the Association of the Bar of the City of New York, Vol. 13, *6*, 396.

2. *John C. Hogan*, A Selective Bibliography on the Legal and Political Aspects of Space, The Rand Corporation, 1958, also published in *Saint Louis University Law Journal*, Vol. 5, *1*, 108—133 (Spring, 1958).

3. *John C. Hogan*, Space Law Bibliography, The Journal of Air Law and Commerce, Vol. 23, *3*, 317—325 (Summer, 1956).

4. *Yevgeny A. Korovine*, Bibliography, International-Legal Questions on the Mastery of Cosmic Space (June, 1960).

5. *Martin Menter*, Astronautical Law, pp. 73—84 (Industrial College of the Armed Forces, Washington D.C., 1959).

6. *Eugène Pépin*, Bibliographie, Les Problèmes Juridique de l'Espace, La Revue Française de Droit Aérien, *4*, 24—46 (Sirey, Paris, 1959).

7. *Michel S. Smirnoff*, Jugoslovenska Bibliografija Vazduhoplovnog Prava (Institut Za Medunarodnu Politiku I Privredu, Belgrade, 1959).

8. United Nations, A Bibliography of the Law of Outer Space — Preliminary Edition (United Nations Library, New York, N.Y., 1958).

9. United States Department of State, Social Science Research on Outer Space, A Selective Listing, pp. 9—15 (External Research Division, Bureau of Intelligence and Research, Washington, D.C., 1959).

10. United States House of Representatives, Bibliography of Space Law, Staff Report of the Select Committee on Astronautics and Space Exploration, pp. 38—60 (United States Government Printing Office, Washington, D.C., 1959).

11. University of Oklahoma, Bibliography of the Space Law Collection (Law Library, Norman, Oklahoma, 1959).

12. United States Department of the Air Force, Space Law — The Legal Aspect, Special Bibliography *No. 161* (Air University Library, 1958).

Appendix 5

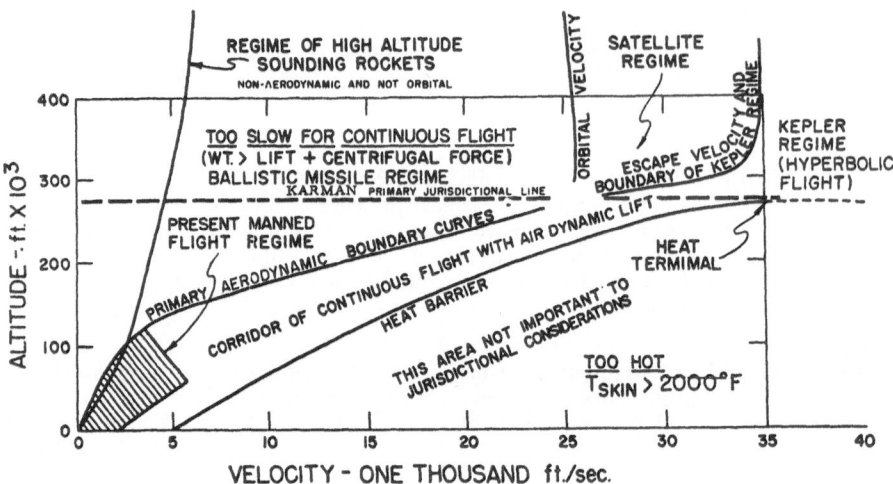

Diagram showing regimes of atmospheric and extra-atmospheric flight and depicting the jurisdictional boundary lines.

Appendix 6

AIRSPACE JURISDICTION — THE STATUTORY LAW OF THE NATIONS OF THE WORLD. Statement of *Andrew G. Haley.*

The launching of the first Sputnik was not only one of the world's most impressive scientific events — it was also the advent of a new epoch involving basic extensions of man's concepts of the social sciences. What actually was the legal status of Sputnik? As Sputnik's orbit lay outside the airspace of the earth, it was obvious to man that no jurisdictional objection could be made by any signatory to the Chicago Convention, or indeed by any nation, signatory of not, in the world community, pursuant to any international treaty or any principle of international law.

At first only idle curiosity led me to wonder whether the statutory law of any nation might be violated — whether any legislature by prophetic foresight had enacted laws making claim to sovereignty in outer space. I spent weeks after the launching of Sputnik I studying the statutory laws of many nations. My research was limited by language barriers. I searched through every compendium in existence and found no complete answer. Finally, I sought the aid of the Chairman of the Senate Committee on Interstate and Foreign Commerce of the United States, the Honorable Warren G. Magnuson. In November of 1957, Senator Magnuson induced the Library of Congress to undertake translations covering the statutory language outlining jurisdiction over airspace of all the nations of the world. On December 3, 1957, the first report, through Senator Magnuson, on the statutes of France, USSR, Western Germany and Sweden, was received. A week later I received reports on the Latin American nations, Spain, Portugal and the Philippines. On December 17, 1957 reports came to me outlining the jurisdiction over airspace of Bulgaria, Poland, Romania, Belgium and Greece; and a week later reports covering Czechoslovakia, Luxembourg and Yugoslavia arrived. Then in quick succession came reports on Denmark, Finland, Norway and Italy, followed by Australia, Burma, Canada, Ceylon, India, Ireland, Israel, New Zealand, Pakistan and the Union of South Africa.

Then I received reports on Iran, Egypt, Albania, Afghanistan, Iceland, Ethiopia, Jordan, Lebanon, Saudi Arabia, Libya, Liberia, Iraq, Turkey, and Syria.

In coordinating this material Mr. Lawrence Keitt, Law Librarian of the Library of Congress, made this notable statement:

"The origin of the term 'airspace', as used in the 1944 Convention, goes back to the Paris Convention of 1919 relating to the Regulation of Air Navigation. In the French text of this Convention, the term 'espace atmospherique' (atmospheric space) is used. The term 'airspace' of the Chicago Convention corresponds to the term 'espace atmospherique' of the Paris Convention. According to C. Wilfred Jenks in his article 'International Law and Activities in the Space', *International and Comparative Law Quarterly* (1956), pages 99—119, international air law recognizes the sovereignty of a state as extending only to the atmospheric space above its territory."

AFGHANISTAN.

The statutory language outlining jurisdiction over airspace in Afghanistan is contained in the "Civil Aviation Act of 1956" which was passed in March 1956. This law created the Civil Aviation Council which is charged with the supervision of the organization and administration of civil aviation. The text of this law is contained in the "Civil Aviation Act of 1956", published by the Government of Afghanistan, Kabul, 1956 (English and Persian texts), 6 pages. This publication was not, however, available for consultation. It is learned from a secondary source that the law includes a provision on Afghanistan's sovereignty over airspace over its territory. Furthermore, the Civil Aviation Act of 1956 of Afghanistan is said to have a specific reference to the Convention on International Civil Aviation, signed in Chicago on December 7, 1944. The scope of the Afghanistan Civil Aviation Act apparently is limited to air navigation matters and does not cover air transport matters.

ALBANIA.

1. *Before Communism.* Albania did not adhere to the Convention of Aerial Navigation signed in Paris in 1919, nor to the Warsaw Convention of 1929.

Up to the Italian occupation of Albania in 1939, apparently no legislation had been passed in Albania with regard to airspace. However, flying over Albania, with the intention of eventually landing somewhere within the national territory, could be done only pursuant to permissions granted in advance by the Albanian Foreign Ministry in accord with the Ministry of Interior (*see Fletorja Zyrtare,* No. 9, 1928).

2. *Present Situation.* a) *Legislation.* Thus far it does not appear that the People's Republic of Albania has enacted any legislation outlining jurisdiction over "airspace". Decree No. 1553, *Gazeta Zyrtare,* No. 17, 1952 defines at length the territorial sovereignty of the Republic, establishes border lines on land and water and directs the Armed Forces of the Republic to defend, as a whole, the national integrity of the country, but is does not mention "airspace" and its eventual limits. And Sec. 260 of the Criminal Code of the People's Republic of Albania provides penalties for those who break rules on international flights, and adds that law shall punish "especially those who enter or leave the People's Republic without having flight permission" in which the route of navigation, landing areas and altitude must be specified.

Albania's claim to sovereignty over "airspace" may be implied from the continual (since 1949) notes of protest of the Ministry of Foreign Affairs of the People's Republic of Albania to the United Nations against the governments of the neighboring countries (Italy, Greece and Yugoslavia) for violating the airspace of the Republic. These notes have been regularly published in the main official organs of the communist press of Albania such as *Zeri i Popullit,* organ of the Central Committee of the Communist Party, *e.g.* on July 18, November 25, 1951, and March 15, 1952. These notes of protest as a rule are not explicit as to the precise legislature of the violation as they merely emphasize that "the space above the territory of the Republic has been violated".

b) *Treaties.* The present Albanian regime did not join the International Civil Aviation Convention signed in Chicago in 1944, nor is there any material to show that Albania entered into any bilateral or multilateral agreement with the other Eastern and Central Europe communist-dominated countries which touch upon this subject.

On December 7, 1955, Radio Tirana broadcast that, as a result of talks between Albanian and Soviet delegations regarding air traffic, an agreement and related protocol was signed in Tirana on December 6, 1955, between the two countries. However, the contents of this agreement have not been published in the Albanian official gazette.

AUSTRALIA.

The national legislation relating to air navigation for the nations included in this report encompasses all provisions bearing on jurisdiction over airspace. In most cases these provisions are of a rather general nature and grant authority to a Governor General or Cabinet Minister to prohibit or restrict flying over specified areas.

According to Baalman, *Outline of law in Australia,* p. 237 (Sydney, The Law Book Company of Australasia Pty, Ltd., 1955), the common law rule *cujus est solum ejus est usque ad coelum* (whoever owns the soil owns all that lies above it) prevails in Australia.

Authority to control airspace in Australia is contained in: Air Navigation Act, 1920 —1950, in *Acts of the Parliament of the Commonwealth of Australia, 1901—1950, 1,* 157—58. Section 5. — (1) reads:

> "The Governor General may make regulations —
>> (a) for the purpose of carrying out and giving effect to the Chicago Convention . . .
>> (b) prescribing all matters
>>> (i) in respect of air navigation which are necessary or convenient to be prescribed in relation to any matter with respect to which the Parliament has power to make laws; or
>>> (ii) which are necessary or convenient to be prescribed in respect of air navigation within any territory of the Commonwealth or to or from any such Territory."

It should be noted that the common law rule referred to has been altered in the State of New South Wales by the Damage by Aircraft Act, 1952, in *Statutes of New South Wales, 1952,* Act No. 46, which excludes all general liability for trespass by aircraft in flight but which makes the owner liable for any actual damage caused by his aircraft, without need to prove negligence.

BELGIUM.

Belgium has adopted the theory of complete sovereignty over the airspace above its territory by ratifying the two international conventions which provide for exclusive jurisdiction over the superincumbent airspace.

1) Convention relating to the regulation of aerial navigation signed at Paris October 13, 1919 (11 League of Nations Treaty Series 173), ratified by Belgium on August 16, 1922 (*Moniteur Belge,* November 16, 1922).

> *Art. 1.* The High Contracting Parties recognize that every Power has complete and exclusive sovereignty over the airspace above its territory.

2) Convention on International Civil Aviation signed at Chicago December 7, 1944 (US Treaties and other international Acts 1591, 1947), ratified by Belgum on April 30, 1947 (*Moniteur Belge,* December 2, 1948).

> *Art. 1.* The Contracting States recognize that every State has complete and exclusive sovereignty over the airspace above the territory.

BULGARIA.

I. Bulgarian legislation now in force reflects the theory that the People's Republic of Bulgaria has complete and exclusive sovereignty over the airspace above its territory, *i.e.,* over its land areas and the territorial waters adjacent thereto.

The principle that Bulgaria claims sovereignty over the airspace above its territory may be deduced from the Edict on Territorial and Inland Waters of October 23, 1952 (IPNS No. 85, as amended, November 9, 1951, IPNS No. 90).

> *Sec. 5.* The inland and territorial waters of the People's Republic, as also the airspace above them and the sea bed and subsoil covered by them, shall be part of the territory of the People's Republic and subject to the laws of the Republic alone.

Sec. 6. The People's Republic of Bulgaria shall exercise its sovereignty over the territorial waters as specified in Section 5 by virtue of the existing laws, rules of international law, and treaties and agreements concluded with other states.

Legal writers of present day Bulgaria made the following statements. Todor Gabrovski, *Vezdushnoto pravo na NR Bulgaria* (The Air Law of the People's Republic of Bulgaria). *Sotsialistichesko pravo* (Sofia) 1956, 5, 49.

"The People's Republic of Bulgaria has complete and exclusive sovereignty over the airspace in our country ..."

Petko Stainov and A. Angelov, *Administrativno pravo na NR Bulgaria, spetsialna chast* (The administrative Law of the People's Republic of Bulgaria, Special Part). Sofia, 1954, p. 210.

"Despite the attempts of some (above all of Americans) authors to develop imperialistic views concerning the freedom of transit in the airspace for all States, in international law [the rule] seems to be established that the States reserve in every respect the sovereignty over the airspace above their territory ... And our legislation in this field is based on this view ..."

II. At present, Bulgaria is a party to only one multilateral international aviation convention: Convention for the Uniformilization of Certain Rules Regarding Air Transportation, signed at Warsaw, October 12, 1929. Bulgaria joined this convention on September 23, 1949 (Todor Gabrovski, *op. cit.,* p. 49).

The present Bulgarian Government did not join the International Civil Aviation Convention of December 7, 1944, signed at Chicago, which superseded the Convention for the Regulation of Aerial Navigation, adopted at Paris, on October 13, 1919 (The Paris Convention of 1919 was joined by the previous Bulgarian Government; DV No. 8, April 14, 1931).

Neither did the present government participate in or join the Geneva Air Convention of 1948, nor the Conference at Rome of 1952.

However, the Government of the People's Republic of Bulgaria entered into a number of bilateral agreements and conventions on air communication with other Eastern and Central European Countries, including the Soviet Union:

(a) Romania, on July 22, 1947, at Sofia
(b) Yugoslavia, on August 2, 1947, at Belgrade
(c) Czechoslovakia, on January 22, 1948, at Sofia
(d) Poland, on May 16, 1949, at Warsaw
(e) Hungary, on June 1, 1949, at Budapest
(f) USSR, on March 5, 1955, at Moscow
(g) East Germany, on July 30, 1955, at Sofia.

After the reconciliation between Yugoslavia (expelled from the Cominform in 1948) and the Soviet block, Bulgaria and Yugoslavia entered into a second air communication convention, October 1, 1955, at Belgrade.

On January 14, 1956 another agreement on air communications was signed between the Bulgarian TABSO (civil aviation agency) and the Yugoslav YAT (air transport agency).

A Convention for Technical Cooperation in the Field of Civil Aviation, was signed on August 4, 1956 at Sofia by representatives of Bulgaria and the Soviet Union.

None of the international conventions and agreements to which the present Bulgarian government is a party contains provisions outlining the jurisdiction over "airspace" or any definitions concerning the extent of the airspace included within the claim of sovereignty.

Note: The International Convention for the Regulation of Aerial Navigation, held at Paris October 13, 1919 (17 Supplement AJIL 1923, p. 198) to which the previous Bulgarian Government was a party contained the following provision:

Art. 1. The high contracting parties recognize that every Power has complete and exclusive sovereignty over the airspace above its territory.

For the purpose of the present convention the territory of a State shall be understood as including the national territory, both that of the mother country and of the colonies, and the territorial waters adjacent thereto.

BURMA.

The Burma Aircraft Act in the *Burma Code, 1944,* vol. 2, 526 ff. provides:

"The Governor [1] may, by notification in the *Gazette,* make rules, regulating ...
"5(1) the prohibition of flight by aircraft over any specified area either absolutely or at specified times or subject to specified conditions and exceptions; ...

"6(1) If the Governor is of opinion that in the interests of public safety or tranquillity the issue of all or any of the following orders, is expedient, he may, by notification in the *Gazette* ...
(b) prohibit, either absolutely or subject to such conditions as he may think fit to specify in the order, or regulate in such manner as may be contained in the order, the flight of all or any aircraft or class of aircraft over the whole or any portion of Burma."

CANADA.

The Aeronautical Act of Canada in *Revised Statutes of Canada, 1925,* vol. 1, chapter 2, states:

"4(1) Subject to the approval of the Governor in Council, the Minister may make regulations to control and regulate air navigation over Canada and the territorial waters of Canada and the conditions under which aircraft registered in Canada may be operated over the high seas or any territory not within Canada and without restricting the generality of the foregoing, may make regulations with respect to ...
(f) the prohibition of navigation of aircraft over such areas as may be prescribed, either at all times or at such times or on such occasions only as may be specified in the regulation and either absolutely or subject to such exceptions or conditions as may be so specified."

CEYLON.

The Air Navigation Act, No. 15 of 1950 in *The Acts of Ceylon, 1950,* provides in Article 2 that ...

"Without prejudice to the generality of the powers conferred by subsection (1) regulations may be made under this Act for or with respect to all or any of the following matters ...
(m) the prohibition of the flying of aircraft over such areas in Ceylon as may be specified in the regulations."

Article 7 of the same act states:

"In time of war, whether actual or imminent, or of great national emergency, the Minister of Defense and External Affairs, may by general or special order regulate or prohibit, either absolutely or subject to such conditions as may be contained in the order, the navigation of all or any descriptions of aircraft over Ceylon or any portion thereof, or the territorial waters adjacent thereto."

CZECHOSLOVAKIA.

In Czechoslovakia a new Law on Civil Aviation was enacted on September 24, 1956 (Law No. 47, 1956 Coll.). The sovereignty of the State over the airspace is stated in its Section 3 as follows:

Sec. 3. Sovereignty over the Airspace
The Czechoslovak State has complete and exclusive sovereignty over the airspace above its state territory.

This provision restates the principle expressed in Article 1 of the Convention on International Civil Aviation of December 7, 1944, of which Czechoslovakia is a contracting party.

There is no definition of airspace in Law No. 47, 1956 or in any other Czechoslovak statutes as far as they are available in this Library.

In a study published in 1956 two Czechoslovak lawyers discussed sovereignty over airspace as follows:

> At the time when the principle of sovereignty over airspace was defined there was no need to consider jurisdiction over the stratosphere because of the technical developments of aviation. The possibility of flights into the stratosphere and of the use of the stratosphere became a reality only recently. In dealing with this question the danger to a country from the air must first be taken into account and [also] the right of each country to defend itself against possible attack. The fact that technical developments made it possible to reach the stratosphere obviously did not eliminate the danger of sudden attack; on the contrary, the possibility of such attack and of a surprise [attack] was increased. If the rules of international law which declare that the sovereign rights of a State extend to the airspace, have to fulfull their functions under conditions of the vast technical development of aviation, they must be applicable also to the stratosphere [2].

The article concludes:

> The launching of an artificial satellite will be an historical event in the history of world science and will be made, in fact, with the consent of the states of the world, given implicity or expressly. Such consent, obviously, will not be a denial of the sovereignty of states over the airspace above their territory. If it appears that the launching of an artificial satellite may present obvious danger to states and that this invention could be misused to endanger the security of states, it is self-evident that states may protest to the international agencies invoking the principle of sovereignty (over the airspace) [3].

The principle of the sovereignty of a state over its airspace, stated in Section 3 of Law No. 47, 1956, is implemented in Sections 47, 49 and 52 of this law.

Sec. 47. Scheduled international transportation by foreign civil aircraft

(1) The civil aircraft which are not registered in the Czechoslovak register of aviation (here after referred to as foreign aircraft) and which are used for scheduled international transportation, may cross the Czechoslovak state borders only if the operation of a scheduled foreign service of international transportation using foreign aircraft has been expressly authorized in accordance with the aviation agreement between the states.

(2) The permit referred to in Subsection 1 shall be granted by the Ministry of Transportation in agreement with the Ministry of Foreign Affairs.

(3) The permit shall also set forth the following conditions; a) sector of flight, *i.e.,* the sector within which the foreign aircraft must cross the state border of the Czechoslovak Republic when entering and departing and its altitude while crossing the state border; b) the route of flight (Sec. 31, Subsec. 2) and the place of compulsory and permissible landing; c) the period of validity of the permit.

(4) If an aviation agreement has not yet been negotiated between the states, the Ministry of Transportation, acting in agreement with the Ministry of Foreign Affairs, may grant a temporary permit for scheduled operation to a foreign service engaged in international air transportation.

(5) The permits granted shall be entered in the Czechoslovak aviation register.

Sec. 49. Compulsory landing of a civil aircraft engaged in an international flight under extraordinary circumstances

(1) If a civil aircraft in distress, or for some other reason, crosses the state border of the Czechoslovak Republic outside the fixed air sector or deviates from the prescribed or approved route it must land without delay as soon as it becomes aware of this or receives a signal to land, at a designated place or, if one is not designated, at

the nearest aerodrome on the territory of the Czechoslovak Republic. The landing of the aircraft must be reported to the nearest security agency without delay.

(2) If the aircraft fails to obey the signal to land it may be forced to land after it has neglected a second signal.

(3) A civil aircraft which lands under the conditions stated in the preceding subsections, may resume flight only with the consent of the Ministry of Transportation acting in agreement with the Ministry of the Interior.

Sec. 52. Non-scheduled international flights of foreign aircraft
Non-scheduled international flights of foreign aircraft over the state border of the Czechoslovak Republic may be made only with the consent of the Ministry of Transportation acting in agreement with the Ministry of Foreign Affairs. In such cases the Ministry of Transportation may impose on foreign aircraft the obligation to land on the territory of the Czechoslovak Republic, or impose additional limitations upon them.

Czechoslovakia has concluded bilateral agreements concerning aviation with the following countries:

Belgium (January 25, 1937)
Bulgaria (January 22, 1948 with Protocol of August 5, 1955)
Denmark (May 14, 1947 and Exchange of Notes of October 1, 1949 and January 14, 1950)
Finland (July 13, 1950)
France (July 27, 1946 and Exchange of Notes of September 1 and October 7, 1948)
East Germany (August 8, 1955)
Hungary (June 9, 1947)
Ireland (January 29, 1947)
Netherlands (September 1, 1947)
Norway (May 7, 1948)
Poland (January 24, 1946)
Romania (September 13, 1946 with Protocol of August 4, 1948)
USSR (February 26, 1955)
United States of America (January 3, 1946)
Sweden (October 15, 1947 and Exchange of Notes of November 5 and December 22, 1949)
Switzerland (September 10, 1947)
Turkey (March 5, 1947)
Yugoslavia (March 14, 1948)

As far as the text of these treaties is available in the Library none of them contains any definitions relating to airspace. They are agreements on the establishment of international aviation routes and the operation of international aviation services between the contracting parties.

DENMARK.
The only Danish statute which explicitly declares that airspace over Danish territory is subject to Denmark's sovereignty is the Royal Decree relating to the Entry During Peace Time of Foreign Warships and Military Aircraft into Danish Territory [4]. Section 2 of this decree provides:

"Danish territory in these regulations shall mean the Danish land territory and the Danish waters together with the airspace above it."

In accord with this are also the views of Danish writers who maintain that there is no definite upper limit to sovereignty in airspace [5]. Also in accord with this principle is the Danish Aviation Act of May 1, 1923 [6], which provides in Section 2 that "Air traffic within Danish territory is permitted only according to the provisions of the present Law".

The Aviation Act further provides (Sec. 3) that only aircraft of Danish nationality and those foreign aircraft which meet certain requirements are admitted to the airspace above Danish territory. Danish nationality is possessed by aircraft which are entered in the official Danish Aircraft Register and display the required identification marks (Sec. 4) [7].

Section 6, *lex cit.* provides which aircraft may be entered in the register:

> "An aircraft may be registered in the Realm only when it has Danish owners. The Danish State, Danish local government units, associations, societies, corporations, and foundations are considered Danish owners. If the owner is a joint stock company, the company's main office must be located in Denmark and the members of the board must be shareholders, Danish nationals and domiciled in the Realm. The Ministry for Public Works may waive the last-mentioned condition but not in regard to the president or for more than one-third of the other members of the board.
>
> "An aircraft which possesses the nationality of another state may not be registered in the Realm."

It is further required for registration that the aircraft must have as its regular home field an approved airport in Denmark, present proof of airworthiness and show that that it is covered by insurance or security of some other kind (Sec. 7).

Section 14 of the Aviation Act lists the conditions under which an aircraft not of Danish nationality is allowed to fly over Danish territory:

> "(a) If the craft is in possession of the nationality and registration certificate or a corresponding certificate issued by a public authority of a foreign state with which an agreement has been concluded to the effect that the craft domiciled in such a state shall be entitled to such a flight; or
>
> "(b) if the Ministry of War or the Ministry of the Navy, as far as foreign military aircraft are concerned, has issued a permit for such a flight; or
>
> "(c) if the Ministry of Public Works, in regard to foreign civil aircraft, has issued such a permit upon application from person on whose account the craft isued."

Further exception is made for test flights for various purposes of airworthiness, *etc.*, by aircraft not otherwise qualified for a flight permit (Sec. 15).

The provisions of the Aviation Act are applicable also to free balloons, airships, airplanes, gyroplanes, helicopters and ornithopters, which have a pilot aboard and which are intended for the transportation of persons or goods [8]. Section 1 of the Act also provides that the use of aircraft operated without a pilot aboard is subject to special rules. These rules for the use of pilotless aircraft, gliders and parachutes have been issued by the Minister for Public Works on December 22, 1956 [9]. Under these regulations, such pilotless aircraft or similar devices may be used only under a permit issued for each case by the Aviation Directorate.

Foreign military aircraft may come into Danish airspace only after a proper clearance has been effected through diplomatic channels. Exceptions to this requirement are aircraft in distress and those aboard a warship [10].

Denmark is also party to the International Convention on Civil Aviation of 1944 which declares that every state has exclusive jurisdiction over the airspace above its territory [11].

Denmark's land territory bounding on Germany is determined by international treaties [12].

Danish territorial waters, as a rule, include a belt of 3 miles measured from the outermost islands or islets which are not submerged under water [13]. The same three mile rule applies in regard to Greenland [14]. For special purposes, like customs control, Denmark claims jurisdiction over waters extending out 4 miles, but it does not consider the additional one mile belt as its territory [15].

EGYPT.

Egypt, in general, claims jurisdiction over the airspace above its territory. Section 1 of the Décret-Loi No. 57 on Air Navigation of May 23, 1935 (*Journal Officiel* 47) provides:

Sec. 1. The State exercises complete and exclusive sovereignty over the airspace above its territory. The term "territory" includes the adjacent territorial waters [16].

This provision restates the principle expressed in Article 1 of the Chicago Convention on International Civil Aviation of December 7, 1944 of which Egypt is a contracting party [17].

ETHIOPIA.

A recent survey [18] states that "Ethiopia has neither law nor regulation on aviation. The efficient ICAO Technical Assistance Mission concentrated its efforts on the technical aspects of aviation. However, although the mission acted without benefit of a lawyer, a short draft of basic law was prepared and submitted to the Government as early as 1950. After translation in Amharic, the draft is still before the Legislative Council."

This information, however, does not appear to be entirely correct in the light of another source. The most authoritative book on Ethiopian laws compiled by the Advocate General and General Adviser to the Imperial Ethiopian Government, [19], although it does not deal specifically with the subject, refers to "all applicable aeronautical regulations issued by the Imperial Ethiopian Government Department of Civil Aviation" to which the Imperial Ethiopian Aero Club and its member shall be subject [20]. It thus would seem that the matter has been regulated to some extent, although the pertinent sources were not available for the purposes of this study. Furthermore, there are also special regulations pertaining to the Ethiopian Air lines [21].

FINLAND.

In addition to being a party to the International Convention of Civil Aviation of December 7, 1944 [22], which affirms the principle that every state has exclusive jurisdiction over airspace above its territory, Finland has also based its own legislation concerning sovereignty over airspace on the sovereignty doctrine [23]. The Finnish general law on aviation, the Aviation Act (*Ilmailulaki*) of May 25, 1923, as amended [24], provides in Section 1 that air traffic within and over Finnish territory is permitted only under the rules and regulations of the law and furthermore, in Section 2, that only aircraft of Finnish nationality and those which have qualified under other provisions of the law shall have the privilege of flying over Finnish territory. The Finnish claim of sovereignty over airspace above its territory is explicitly stated in Section 2 of the Regulation relating to the Movement During Peacetime of Foreign Warships, Merchant Vessels and Aircraft of April 28, 1938 [25]. This provision states:

> "Land and sea territory within the borders of the state and the airspace above it is considered Finnish territory."

The same regulation in Sec. 21 provides also that a foreign aircraft may fly over Finnish territory only in accordance with international agreements or other regulations or by a special permit. Excepted from such formalities are aircraft carrying the head of a foreign state and aircraft accompanying such a plane (Sec. 22). However, in such cases notice through diplomatic channels must be given (Sec. 7).

Provisions regarding Finnish nationality of aircraft and foreign aircraft entitled to fly over Finnish territory are contained in Secs. 3—5 of the Aviation Act. Sec. 3 of this Act provides:

> "An aircraft possesses Finnish nationality if it is entered in the Finnish public aircraft register and displays the nationality and registration symbols or if it belongs to the government or has been surrendered for the latter's exclusive use."

Sec. 4 contains the requirements for entry into the aircraft register:

> "An aircraft may be entered in the register only if it is Finnish-owned. The Finnish state, Finnish nationals and domestic companies, societies, other associations, and foundations are considered Finnish owners. If the owner is a company, society, other association, or foundation, the members of its board must be Finnish nationals domiciled in this state who are shareholders in the company or members (partners) of the society or association . . ."

The flying of foreign aircraft over Finnish territory is regulated as follows in Sec. 5:

> "An aircraft which is not of Finnish nationality may fly over Finnish territory if
> (a) the craft possesses a nationality and registration certificate or a corresponding certificate from a public authority of a country which is party to an agreement under which the aircraft of such a country have the right to fly over Finnish territory,
> (b) the craft has been issued a special air navigation permit."

As mentioned above, Finnish territory includes the airspace both over the land and sea territory. The land territory of Finland has been fixed by agreements with neighboring countries (Soviet Union, Sweden, Norway). Finnish territorial waters are determined in the Law on the Limits of Finnish Territorial Waters (*Laki Suomen aluevesien rajoista*) of August 18, 1956 [26] and the regulation of the same date for its application [27]. Section 5 of this law provides that Finnish territorial waters extend, as a general rule, four nautical miles (7408 m) from the outer border of coastal waters which is formed by the surf line. As an exception, territorial waters extend only to three nautical miles from outlying islands, reefs or rocks (*lex cit.* Sec. 6).

According to Sec. 2 of the Regulation relating to the Movement during Peace Time of Foreign Warships, Merchant Vessels and Aircraft, Finnish sovereignty extends to the airspace (air column) above its sea and land territory. Neither this regulation nor other statutory provisions set any limit to the sovereignty in the air. This seems to suggest that the "airspace" in its entirety is subject to the jurisdiction of Finland, but not the "airfree space".

Sec. 2 of the Aviation Act defines as aircraft such airplanes, dirigible airships and free balloons which may be used as a means of conveyance. These terms are not elaborated on in the regulations and thus conceivably could include also unmanned airplanes and rocket propelled missiles.

FRANCE.

The following legal provisions on airspace have been found in French statutes, or agreements to which France is a party:

1. *Art. 552* of the Civil Code:
   Ownership of the land includes ownership of what is above and below it.
2. Convention Relating to the Regulation of Aerial Navigation, signed at Paris, October 13, 1919, with additional Protocol signed at Paris May 1, 1920 (11 League of Nations Treaty Series 190, 1922):
   Art. I, Sec. 1. The High Contracting Parties recognize that every Power has complete and exclusive sovereignty over the airspace above its territory.
3. Chicago Convention on International Civil Aviation, 1944; (US Treaties and Other International Act Series 1591. Department of State, 1947):
   Art. I: The contracting States recognize that every state has complete and exclusive sovereignty over the airspace above its territory.
4. Code of Civil and Commercial Aviation; Decree No. 55-1590 of November 30, 1955 (*Journal Officiel*, Dec. 6, 1955):
   Art. 17. Aircraft may fly freely over French territories. However, foreign aircraft shall not fly over French territories unless such right is granted by a diplomatic convention or if they obtain, for this purpose, an authorization which shall be special and temporary.

GERMANY.

Air sovereignty was restored to Germany by the Convention on the Settlement of Matters Arising out of the War and the Occupation [28] which was concluded between the Western Powers and the Federal Republic of Germany at Paris on October 23, 1954, taking effect May 5, 1955.

The basic legislation governing aviation in the Federal Republic of Germany [29] is the Law Concerning Air Traffic in the version of the Law of August 21, 1936 [30] as amended by the Laws of September 27, 1938 [31], and of January 26, 1943 [32], and the Decree issued for its implementation of August 21, 1936 [33] as amended by the Decrees of March 31, 1937, July 12, 1937, December 15, 1937 [34], September 30, 1938 [35], November 5, 1954 [36], and of June 21, 1955 [37].

The laws cited were searched but no definitions of the terms "airspace" and "atmosphere" were found.

The translation [38] of several provisions of the Law concerning Air Traffic and the Decree concerning Air Traffic which may be of interest to the inquirer follow.

LAW CONCERNING AIR TRAFFIC [39]

Sec. 1 (1) The use of the airspace by aircraft shall be free insofar as it is not restricted by the present law and regulations issued for its enforcement.

Sec. 1 (2) Aircraft within the meaning of the present law are airplanes, airships, gliders, balloons, kites and similar implements designed for motion in the airspace.

It may be pointed out in passing that the provision stated in Sec. 1, Subsec. 1 affects private rights considerably, since it vitiates the private law maxim *"cuius est solum ejus est usque ad coelum"* in effect under the German Civil Code [40]:

Sec. 905. The right of the owner of a piece of land extends to the space above the surface and to the substance of the earth beneath the surface. The owner may not, however, forbid interference which takes place at such a height or depth that he has no interest in its prevention.

Sec. 12. Certain areas may be closed temporarily or permanently to air traffic either entirely or below a certain minimum altitude (restricted air areas).

DECREE CONCERNING AIR TRAFFIC [41]

Sec. 1. Aircraft within the meaning of the present decree are airplanes, airships, gliders, free and captive ballons, parachutes which serve for jumping purposes, as well as kites and model aircraft weighing more than 5 kg.

The principle that Germany claims sovereignty over the airspace above its territory may be also deduced from Section 102 of that Decree. Sec. 102 reads in part:

"(1) Foreign aircraft may only fly into areas under German sovereignty and travel therein, insofar as this is generally permitted by an agreement concerning air traffic concluded between their country of origin and the German Reich, or when the Reichminister for Aviation has granted special permission to enter."

In addition, the Basic Law for the Federal Republic of Germany promulgated May 23, 1949, effective March 15, 1955, was consulted. The Constitution is silent on the points under investigation. The only provision contained therein dealing with air legislation is Article 73, Point 6 which reads [42]:

"The Federation shall have exclusive legislation on:

— — —

6. Federal railway and air traffic.

— — —

Professor A. Meyer, an authority on Air Law and Director of the Institute of Air Law at the University of Cologne, expresses in a recent study [43] the view that the air space over a nation is an integral part of the nation's territory:

"... The Federal Government and the individual States have *de jure* "concurrent control" in the air space above their territories. This follows from the principle that the air space over a nation is an integral part of the nation's territory, and has so been by nature from the earliest days. The different areas of the earth are not to be considered as a plane surface, but can only be conceived three-dimensionally as mankind is also a three-dimensional being. Therefore, the argument that the air space did not become part of a national territory until nations had the physical ability to fly in it and to control it, is, in my opinion, erroneous. The sovereignty of the States in the air space over their land and waters is a compulsory consequence of the natural connection between the air space and the land and water territories below it, though this may have appeared to us as a reality only since air navigation began to develop."

GREECE.

The air law in Greece is contained in international conventions which the Greek government signed; among them the Warsaw Convention of October 12, 1929, ratified by Law No. 596 of May 29, 1937.

No definition of airspace was found in Greek law.

ICELAND.

Iceland has ratified the Chicago Convention on International Civil Aviation of December 7, 1944 [44] and accordingly adheres to the sovereignty principle expressed in Sec. 1 of the Convention as follows:

"The contracting States recognize that every State has complete and exclusive sovereignty over the airspace above its territory."

The Icelandic Aviation Law (*Lög um loftferdir*) of June 14, 1929, as amended [45], is also based on this principle and asserts that flights by aircraft over Iceland's territory are allowed only according to the provisions of the law. As a rule, only aircraft owned by Icelandic nationals may fly over the territory of Iceland and such aircraft have to be registered in Iceland. (*Lex cit.* Secs. 2 and 3.)

The territory of Iceland includes a belt of sea around Iceland three nautical miles wide [46].

INDIA.

The Indian Aircraft Act, 1934, in Sec. 17, provides:

"No suit shall be brought in any Civil Court in respect of trespass or in respect of nuisance by reason only of flight of aircraft over any property at a height above the ground which having regard to wind, weather and all the circumstances of the case is reasonable, or by reason only of the ordinary incidents of such flight."

IRAN.

Iranian law concerning airspace is based on the principle of unrestricted sovereignty by the State over the airspace above its territory. Iranian territory includes both the land and the territorial waters. This follows from the text of the Iranian legislative enactment concerning air navigation. Sec. 1 of the Decree of the Council of Ministers of August 5, 1938 on Air Navigation [47] reads as follows:

"Any foreign airplane desiring to fly over the territory of Iran or to make a stop on it, unless there exists either a special international treaty or a special agreement on this subject, has to secure in advance permission from the Imperial Government and conform to the regulations of which it has been notified.

The territory of Iran includes also the territorial waters."

Sec. 2 of the same law reads:

"Permission has to be requested at least fifteen days in advance, through diplomatic or consular channels, from the Ministry of Foreign Affairs."

A new law on Civil Aviation was issued July 19, 1948 [48] and later Iran published in Persian and in an English translation the Law of Civil Aviation of 1949 [49]. The text of these laws, however, has not been available for this study.

A recent survey of National legislation since the Chicago convention, in referring to the 1949 Iranian law quoted above, made the two following statements: (1) that the Iranian law is one of the very few existing laws in which the scope of the law is limited to air navigation matters; and (2) that a provision on sovereignty over air space, which was a common provision in pre-war laws, continues to appear in the civil aviation law of Iran [50]. Furthermore, it states that the Civil Aviation Board of Iran is the authority responsible for the organization and administration of civil aviation in Iran [51].

IRAQ.

The basic law governing civil aviation in Iraq is the Air Navigation Law No. 41 of August 6, 1939 [52] which is available also in English [53]. This law, however, does not contain any provision which would define the terms "airspace" or "atmosphere", although it has a number of other legal definitions (aircraft, aerodrome, "Iraqi aircraft", *etc.*). The provisions relating to flights over Iraq territory are as follows:

> *Art. 1.* — The following expressions shall have the meaning set out against them:
> THE MINISTER: The Minister of Defense.
> AIRCRAFT: Includes fixed or free balloons, kites, airships, aeroplanes, flying-boats, gliders and any other aircraft supported by gas lighter than air and having means of propulsion.
> AERODROME: Any area of ground or water (including buildings on Aerodromes, Docks or Slipways used by aircraft) set aside wholly or in part for the landing or take off of aircraft.
> *Art. 2.* — No person shall establish an air transport service in 'Iraq or between 'Iraq and any foreign country or between foreign countries via 'Iraq, whether intended to land in 'Iraq or not, without the consent of the Council of Ministers. The Council of Ministers may give their consent subject to such conditions as they prescribe. The Council of Ministers may withdraw their consent if such conditions are contravened.

In addition, a special Regulation Governing the Control of Aerial Navigation No. 54 of 1939 was issued on August 28, 1939 [54]. The integral text of these implementing regulations, issued on the basis of the Air Navigation Law No. 41 of August 6, 1939, and creating the "Board of Control of Civil Aviation" (Art. 5) is as follows:

> *Regulation Governing the Control of Aerial Navigation No. 54 of 1939*
> After perusal of Articles 5 and 10 of the Aerial Navigation Law No. 41 of 1939, pursuant to the proposal of the Minister of Defense and with the approval of the Council of Ministers, I hereby order the promulgation of the following Regulations: —
> *Art. 1.* The Minister of Defense or his representative shall be authorized to inspect the cargo and passengers of all civil aeroplanes with effect from the date on which the Minister considers that war, or internal or external danger to the Kingdom of Iraq is imminent.
> *Art. 2.* The Minister of Defense or his representative is authorized to assign the heights and routes which all civil aircrafts shall follow, as well as the airports for their landing.
> *Art. 3.* The Minister of Defense or his representative may issue the necessary orders and instructions for the enforcement of the provisions of Articles 1 and 2 hereof, subject to the conditions specified in the International Convention for Aerial Navigation.
> *Art. 4.* The Military authorities shall control civil aviation on the declaration of war, take possession of all aircrafts belonging to hostile States, inspect the cargo and passengers of all aircrafts belonging to neutral States, together with the inspection of Offices and Bureaus of their respective Companies. The Military authorities may also prohibit the flying of neutral aircrafts over 'Iraqi territory.
> *Art. 5.* The Minister of Defense shall form a board to be known as "Board of Control of Civil Aviation" composed of the Director of Civil Aviation, Officers of

the Royal 'Iraqi Air Force and a Police representative provided that this board shall be under the presidency of the Minister of Defense or the person acting on his behalf. *Art. 6.* The Board of Control may appoint supervisors in the air ports and sea-ports to ensure control on civil aviation. This Board shall undertake to publish and notify the civil aircrafts of the necessary precautionery instructions and regulations. These instructions shall include the limits of flying, local inspections, definition and delimitation of prohibited areas, the frequent alteration of same, sealing and seizure of wireless apparatus when civil aircrafts are on the ground, prohibiting the carriage of arms, ammunition, explosives, gases, and the use of cameras, *etc.*
Art. 7. This Regulation shall come into force from the date of its publication in the Official Gazette.
Art. 8. The Ministers of Interior and Defense are charged with the execution of this Regulation.
Made at Baghdad this 13th day of Rajab, 1358, and the 28th day of August, 1939.

IRELAND.

The Air Navigation and Trasport Act, 1946, in *Acts of the Oireachtas, 1946* states in Section 11:

> "Without prejudice to the generality of the powers conferred by Sections 9—10 of this Act, the Minister may by Order made under either of the said sections make provision: . . .
> "(i) appointing any area to be a prohibited area for the purpose of this order."

ISRAEL.

Israel on becoming an independent sovereign state in 1948, retained the air navigation law in effect in Palestine during the period that it was under mandate, by Air Navigation (Amendment), Act, 5710-1950, which is in *Laws of Israel,* Volume 4. The original law is the Air Navigation Act, 1920 (10 & 11 Geo. 5, c. 80) in *Laws of Palestine, 1933,* volume 3, page 2401: It provides that full and absolute sovereignty and rightful jurisdiction of His Majesty extends over the air super-incumbent on all parts of His Majesty's dominions and territorial waters adjacent thereto. It further authorizes in Section 3(m), page 2403: "any steps to be taken for preventing aircraft from flying over prohibited areas and in time of war or national emergency the Secretary of State may order, regulate or subject to such conditions as may be contained in the order . . . any portion thereof . . ."

ITALY.

I. Statutory Provisions

(1) The basic provisions on aviation law are contained in the Law of August 30, 1923 [55], and the Regulation of January 11, 1925 [56].
According to the Law of 1923 the State has complete and exclusive sovereignty over the airspace above its territory, which includes the land areas and territorial waters adjacent thereto under the sovereignty of the State (Sec. 1). However, the law neither defines nor explains what is meant by the phrase "airspace" or "atmosphere".
(2) The Code on Navigation was enacted in 1942 [57]. It provides that the territorial waters of Italy extend six miles from the watermark of the seacoast, the islands, the outer-most points of the port installations and the line of the inland waters (Sec. 2).

II. International Agreements Adopted by Italy

(1) Another important source of the Italian law on aviation is contained in the Convention for the Regulation of Aerial Navigation adopted in Paris on October 13, 1919. This convention was ratified by Italy in 1922 [58]. It remained the basic international law on the subject of aerial navigation until it was superseded by the International Civil Aviation Convention. Questions of sovereignty in the air were passed upon as well as the rights of nationals of one country to fly in or through the airspace of another country. Thus, it provided that every power has complete and exclusive sovereignty over the airspace above its territory (Art. 1).
(2) The International Civil Aviation Convention in Chicago of December 7, 1944 superseded the Convention of Paris of 1919. Italy ratified this convention in 1948 [59]. It

provided that every state has complete and exclusive sovereignty over the airspace above its territory, which includes the land areas and territorial waters adjacent thereto under the sovereignty of the State (Art. 8, Secs. 1 and 2).

JAPAN.

The traditional theory that the sovereign has jurisdiction over the "airspace" or "atmosphere" above its territorial land and territorial waters is generally accepted in Japan. As a matter of fact the fundamental concept of *"cujus est solum, ejus est usque ad coelum et ad inferos"* in common law, is also expressed in the provisions of the Japanese Civil Code. Reference is made to Art. 207 of the Civil Code, which reads:

> "Subject to limitations by laws and ordinances, the ownership of land extends both above and below its surface."

Of course, such private rights may only be asserted against other private individuals. The state never does give up the jurisdiction over its territory.

In the past Japan had been adhered to the *Paris Convention for the Regulation of Air Navigation* (1919), and it had enacted *The Law on Air Navigation* in April 8, 1921. This law consisted of 64 Articles, and dealt with such subjects as registration, licensing of personnel, aerodromes, air navigation and transports, as well as penalties. After World War II, a new law was promulgated, commonly known as the *Civil Aeronautic Law* (No. 231) of July 15, 1952. It was subsequently amended by laws No. 278 of 1952, Nos. 66 and 151 of 1953, and No. 60 of 1954, which was divided into ten chapters (162 Articles): *e.g.* General Provisions; Registration; Safety of aircraft; Airman; Airway, Airdrome and Air navigation and Operation of aircraft; Air transportation, *etc.*, Foreign aircraft; Miscellaneous provisions; Penal provisions, and the Supplementary provisions. In the first article of the said legislation it enunciated its purpose, which states:

> "The purpose of this Law is to promote the development of civil aviation by providing for methods of securing the safety of navigation of aircraft in conformity with the provisions of the convention on Inter-national Civil Aviation and with the standards, practices and procedures adopted as annexes thereto, and by establishing order of enterprises carried on by operating aircraft."

Japan is now a member nation to the International Air Transport Agreement as was adopted by the Chicago Convention in 1944. Under the existing law of Japan, and the obligation inherent thereby as a contracting state of the Convention, the theory of jurisdiction over space above Japanese territory, and the interpretation of the pertinent terminology in Air Navigation, are rather similar to those found in the United States. We may conclude, therefore, that this is one more example of American contribution to the Japanese legal system in the aftermath of World War II.

JORDAN.

The situation with respect to jurisdiction over airspace in Jordan is unclear as the text of the law on civil aviation [60] has not been available, and it could not even be ascertained when such a law had been passed. As to recent developments, a draft law and regulations were prepared in 1952 by a British expert. The draft law and the regulations were translated into Arabic, studied by the authorities, and will probably reach the Parliament in the near future [61].

However, there is no doubt that Jordan, like the rest of the Arabic States, exercises exclusive jurisdiction over airspace above its territory. At least, it has concluded a number of international treaties which have their basis in Jordan's claim to exclusive jurisdiction over airspace above its territory. Thus, Jordan has concluded, on May 20, 1948, a bilateral agreement on air transport services with Turkey [62]. Unfortunately, no information other than on the existence of this bilateral agreement between Jordan and Turkey is available [63].

Likewise, Jordan has concluded a bilateral agreement on regular air transportation services with Egypt. It became effective on April 14, 1953 [64].

Survey of Legal Opinion on Extraterrestrial Jurisdiction

LATINAMERICA, PORTUGAL, SPAIN, AND THE PHILIPPINES.

A thorough search has been made through the constitutions, civil codes, and special statutes, for the twenty-three countries involved in this report, for provisions which might define the sovereignty or jurisdiction claimed in each country over the airspace or atmosphere.

The civil codes cover property rights in general, including public and private property. The civil law concept is noticeable herein insofar as ownership of the atmosphere or airspace over privately owned land is concerned, which follows the "column of air" doctrine of Roman law. The owner of the land, with certain restrictions, owns also the subsoil and the atmosphere over his land. Since this is apparently not the information which is required in this report, no provisions of this type have been included.

In a few Constitutions, the principle of sovereignty over the airspace is enunciated, but it was in the more recent type of administrative legislation, in the form of codes or statutes on air transportation and aviation, that the majority of provisions on national sovereignty were located. The pertinent articles of these bodies of law have been translated on the attached sheet for your convenience.

All of the countries covered in this report, with the exception of Costa Rica and Panama, have also ratified or adhered to the most recent major international agreement in this field, the 1944 Chicago Convention on International Civil Aviation, implying that the provisions of the convention would become law of the land, in general terms. This would include in Articles 1 and 2, which spell out jurisdiction as follows: "The Contracting states recognize that every State has complete and exclusive sovereignty over the airspace above its territory" and that the term "territory" is defined as "land areas and territorial waters adjacent thereto under sovereignty, suzerainty, protection or mandate of such State".

ARGENTINA. Law no. 14.307 of July 15, 1954 (Air Code):

Art. 1. This Code will regulate civil aeronautics in the territory of the Republic and in the air space above it, circumscribed by vertical lines in its perimeter.

Art. 2. For the purpose of this Code, territory includes the territorial waters.

BOLIVIA. Supreme Decree of October 24, 1930:

Considering: That while a law is being drafted for air traffic it is necessary to exercise national sovereignty, over the airspace over our frontiers and jurisdictional waters exclusive of any foreign country:

Art. 1. It is established as aerial patrimony of the nation, the perpendicular column of air which covers the surface of the national territory within the limits of the frontiers, the height being determined by the range of the defensive methods of the country.

BRAZIL. Decree-Law no. 483 of June 8, 1938 (Brazilian Air Code):

Art. 1. The United States of Brazil exercise complete and exclusive jurisdiction over the airspace located above its territory and respective territorial waters.

CHILE. Decree with Force of Law on Air Navigation (*Diario Oficial* of May 30, 1931).

Art. 22. The State will exercise complete and exclusive sovereignty over the atmospheric space existing over its territory and its territorial waters.

COLOMBIA. Decree no. 66 of January 12, 1934:

Art. 1. The atmosphere space which covers the territory and territorial waters of the Republic is part of the territory referred to in Article 4 of the Constitution, and therefore, belongs only to the Nation.

Law no. 89 of May 26, 1938:

Art. 1. "National atmospheric space" is that which covers the territory and territorial waters of the Republic.

COSTA RICA. Constitution. November 8, 1949:

Art. 6. The State exercises complete and exclusive sovereignty over the airspace above its territory and over its territorial waters and continental shelf, in accordance with principles of International Law and with treaties in effect.

69

General law on civil aviation. October 18, 1949:

Art. 1. Costa Rica possesses and exercises complete and exclusive sovereignty in the airspace over the continental and insular territory of the Republic and its territorial waters, allowing, in time of peace, freedom of innocent passage for civil aviation in accordance with international conventions and law. Foreign military planes may not be operated in this space without permission of the Ministry of Public Security.

CUBA. Decree no. 548 of April 21, 1928. (Regulation on Civil Aeronautics on the Territory of the Republic and its Territorial Waters):

Art. 1. The sovereignty which the Republic of Cuba exercises over its territory, as defined by Article 3 of the Constitution, includes as well, completely and exclusively, the atmospheric space over said territory and including in same the territorial waters.

DOMINICAN REPUBLIC. Law no. 1915 of Januari 19, 1949 (On Civil Aeronautics):

Art. 1. Civil, commercial, sport or educational aeronautics over the national territory and in the airspace over its territorial waters and adjacent islands, is regulated by the present Law and the respective Regulations which the Executive Power may issue.

ECUADOR. Constitution. December 31, 1946:

Art. 4. In addition to the continental provinces situated in South America, the national territory embraces the adjacent islands, the Archipelago of Columbus, also known as Galapagos, the territorial seas, the subsoil and the atmosphere above them.

Law on Air Transit of September 12, 1936:

Art. 1. The State exercises complete and exclusive sovereignty over the atmospheric space covering its territory, including its territorial waters.

Decree no. 02 of January 13, 1954:

Art. 1. The supervision and control of civil aeronautics within the territory of Ecuador belongs to the Government, as does the construction, operation and maintenance of the airports of the country with their services and installations.

Art. 3. .... Attributes and duties of the Ministry of Public Works are:

(c) To issue those regulations which it deems necessary for the compliance of the objectives of the Board of Civil Aviation, limiting them whenever possible to the international standards of the International Civil Aviation Organization (ICAO) and other international conventions which have been ratified by the Government of Ecuador.

GUATEMALA. Constitution of February 6, 1956:

Art. 3. National property [dominium] includes the territory, soil, subsoil, territorial waters, continental shelf and air space, and covers the natural resources and wealth which may exist in them, without prejudice to the freedom of navigation, maritime and aerial, in accordance with the law and the provisions of international treaties and agreements.

HONDURAS. Decree no. 121 of March 14, 1950 (Law on Aeronautics):

Art. 1. Honduras has exclusive and absolute sovereignty over the air space corresponding to its territory, the territorial waters and the adjacent islands in both oceans.

MEXICO. Law on General Means of Communication (*Diario Oficial* of February 19, 1940). Book IV (as amended to December 30, 1950):

Art. 306. The air space located over Mexican territory is subject to national sovereignty.

For the purpose of this law the term Mexican territory will include the terrestrial extension of the United States of Mexico, the territorial waters and the adjacent islands on both seas, and Guadalupe Island and Revillagigedo Islands situated in the Pacific Ocean.

NICARAGUA. Constitution. November 1, 1950:
Art. 5. The national territory extends from the Atlantic to the Pacific Ocean and from the Republic of Honduras to the Republic of Costa Rica. It includes, in addition: the adjacent islands, the subsoil, the territorial waters, the continental shelf, the submerged lands, the air space and the stratosphere.

Decree no. 176 of September 19, 1956 (Civil Aviation Code):
Art. 1. The air space located over the territory of the Republic of Nicaragua is subject to the national sovereignty.
For the purpose of this Code, the territory includes the terrestrial extensions and the waters which are under the sovereignty or jurisdiction of the Republic.

PANAMA. Regulation of commercial aviation in the Republic (Decree no. 89):
Art. 6. The term "air space", as used in this regulation, should be understood to mean the space of air superposed vertically upon a determined area.
Art. 7. Private airplanes may operate in the Republic provided that the airplanes, as well as the operators of them, have a license issued by the Commission of Aviation.
Art. 8. The following prerequisites will be exacted, as prior conditions for the granting of permission to operate private airplanes in and through the air space of the Republic: . . . .

Civil code (as amended 1916):
Art. 329. Goods in the public domain are: . . .3. The air.

PERU. Regulation of commercial and civil aviation (approved by Supreme Decree of December 18, 1933):
Art. 1. The Republic of Peru excercises absolute sovereignty over the air space included within the limits of its territory and territorial waters.

SALVADOR. Constitution. September 8, 1950:
Art. 7. The territory of the Republic within its present boundaries is irreducible; it includes the adjacent sea within a distance of two hundred marine miles measured from lowest tide and it embraces the air space above, the subsoil, and the corresponding continental shelf.
Art. 1. The air space located above the Salvadorian territory is subject to national sovereignty.
For the purpose of this law, the term Salvadorian territory includes the terrestrial extension of the Republic of El Salvador, the territorial waters and the adjacent islands.

URUGUAY. Decree no. 1877 of December 3, 1942 (Code on Air Legislation):
Art. 1. The State exercises complete and exclusive sovereignty over the atmospheric space which covers its territory and territorial waters.

VENEZUELA. Constitution April 11, 1953:
Art. 2. The extent of the territorial sea, the contiguous maritime zone, and the air space over which the State exercises its vigilance, shall be determined by law.
Civil Aviation Law of June 13, 1944 (Trans. by Bernardo Flores).
Art. 2. The United States of Venezuela exercises full sovereignty over the air space corresponding to the national territory and to its territorial waters, whatever the altitude.

Law on Civil Aviation of April 12, 1955.
Art. 2. The air space located above Venezuelan territory is subject to National Sovereignty.
For the purpose of this Law Venezuelan territory is understood to mean that determined by Article 2 of the Constitution of the Republic.

PORTUGAL. Constitution. August 1, 1935:

Art. 49. The public domain of the State shall comprise the following: . . .

(5) The aerial strata above the land, beyond such limits at the law fixes in favor of the owner of the surface.

Decree no. 13, 537 of April 27, 1927.

Art. 1. Portuguese or foreign aircraft travelling in the air space which covers the metropolitan and colonial territory of Portugal and the respective territorial waters are required to comply strictly with the present Decree with Force of Law.

SPAIN. Law on the Bases of Air Navigation of December 12, 1947:

Art. 3. (1) The Spanish State has sovereignty over the air space located above its territory — metropolitan, protectorates and colonial — and the territorial waters adjacent to the same.

PHILIPPINES. No provision located defining jurisdiction over air space in Constitution, Civil Code, Law on Civil Aviation (1952).

However, the Philippines have made public adherence to the Convention for Unification of Certain Rules Relating to International Transportation by Air, of 1929 (Proclamation 201, of September 23, 1955), and have ratified the Chicago Convention on International Civil Aviation (ratified in Senate on May 5, 1952, and its Protocol on February 3, 1955).

LEBANON.

The Lebanon Civil Aviation Law of January 11, 1949 was published in the Official Gazette of the Republic of Lebanon, No. 3, Year 98, of January 19, 1949. It thereupon was also published as a separate pamphlet [65]. The statutory language outlining jurisdiction over airspace in Lebanon quoted below shows that Lebanon has adopted the theory of complete sovereignty over the airspace above its territory, which includes the land territory and the territorial waters. The pertinent provisions of the Law concerning Civil Aviation enacted on January 11, 1949, read as follows:

Sec. 1. The state has complete and unrestricted sovereignty over the space above its territory. The space is the area of air that is above the land and sea of the territory of the State.

Sec. 2. No airplane is permitted to fly over Lebanese territory or to land on Lebanon territory before securing in advance a permission from the Minister of Public Works, or which has been granted the right of overflying in accordance with international treaties signed between Lebanon and the country to whom the airplane belongs.

The statutory language outlining jurisdiction over airspace in Lebanon is very closely related to that used by the corresponding statute of Syria. These two neighboring countries — Lebanon and Syria — have extremely similar civil aviation laws [66].

LIBERIA.

A recent survey states that "There is no Aviation Department proper, or law on aviation; for a few months past, regulations have been in preparation in order to bring the International Civil Organization (ICAO) Annexes into force" [67]. Liberia is one of the few remaining states which still has no basic aviation law at all [68].

Certain provisions of law which are designed primarily for the collection of customs may be pertinent to an extent. They are found in the new Liberian Code of Laws of 1956 [69] and read:

1030. Airports of entry defined, established. —

The President is authorized to designate such places as he deems necessary as airports of entry.

The President is further authorized to extend the limits of any seaport of entry to include any airport of entry established under the provisions of this section.

Prior legislation: L. 1941-42, ch. LV, secs. 1, 2 (a), (c).

Cross references: Ports of entry defined, see sec. 853 of the Title. Extent of seaports of entry, see sec. 854 of this Title.

1031. Rules of navigation and Customs Law may be made applicable to air navigation and airports of entry. —

The President is authorized to regulate the application to air navigation of the existing laws and rules and regulations relating to the entry and clearance of vessels and the administration of customs to such extent and upon such conditions as he deems necessary.

Rules and regulations relating to the administration of customs which apply to air navigation and airports of entry shall have the full force and effect of law when issued with the approval of the President.

Violation of such laws, rules, or regulations shall be punished according to the penalties established therein.

Prior legislation: L. 1941-42, ch. LV, secs. 2 (b) in part, 3.

Cross reference: Civil Aviation and Communications Board shall issue rules and regulations controlling civil aviation, see Executive L. sec. 233.

1032. Aircraft from foreign country to land at airports of entry only. —

Except in the case of emergency of forced landing, aircraft entering Liberia from any foreign country shall land at airports of entry only unless permission has been secured from the Civil Aviation and Communications Board to land elsewhere. In the latter case, the owner of such aircraft shall be responsible for the payment of fees and other expense of the official or officials designated to supervise such landing.

LIBYA.

Libya has no special aviation law of its own, except some pre-war Italian air regulations which still appear to be in force. A legislative draft has been approved in 1956, but has not yet become law. The situation is thus summed up in a recent report [70] on national legislations on aviation since the Chicago convention:

Libya. Theoretically under the Constitution, certain pre-war Italian air regulations appear still to be in force. In 1954 an excellent lawyer was sent to Libya as Technical Assistance expert; a preliminary draft was attached to a report, and after several months of discussion with the Libyan officials, a final draft was approved in 1956. However, the law is not yet in force, because of certain difficulties of translation into Arabic. Some regulations have also been prepared by the Technical Assistance Mission; it appears that they are informally observed although not effectively in force, in the absence of a law-giving authority for their issuance.

LUXEMBOURG.

The basic law governing aviation in Luxembourg is the Law of January 31, 1948 concerning the Regulation of Air Navigation which was published in the Official Gazette No. 11 (*Mémorial du Grand-Duché de Luxembourg*) of February 14, 1948, p. 203. This law does not contain any provision which would define the terms "airspace" and "atmosphere".

The provisions relating to flights over Luxembourg territory follow:

Art. 2. The traffic of national aircraft above the territory of the Grand-Duchy shall be free except for restrictions resulting from the present law and those which shall be issued by a Grand-Ducal Decree.

Art. 3. The traffic of foreign aircraft above the territory of the Grand-Duchy shall be subject to authorization by the Minister of Transportation. This authorization shall not be required for the traffic of aircraft registered in countries with which agreements granting reciprocity in this field have been concluded.

Flying over the territory of the Grand-Duchy by foreign military aircraft shall be subject to authorization by the Minister of the Armed Forces.

Art. 4. Flying over the territory of the Grand-Duchy or a part thereof may be prohibited, by Grand-Ducal Decree, to national or foreign aircraft.

Luxembourg adheres to the Convention on International Civil Aviation signed at Chicago on December 7, 1944. This was done by the Law of March 25, 1948, published in the *Mémorial du Grand-Duché de Luxembourg* No. 24 of April 14, 1948, p. 537.

Article 1 of the Convention provides:

"The contracting States recognize that every State has complete and exclusive sovereignty over the airspace above its territory."

NEW ZEALAND.

The Civil Aviation Regulations, 1953, issued pursuant to the Civil Aviation Act, 1948, provide in Article 33:

"(1) For reasons of military necessity, or public safety, the Minister may, by notice in the *Gazette,* proclaim any area of New Zealand territory to be a prohibited or a restricted area for the purposes of these regulations.
"(2) An aircraft shall not —
(a) fly over a prohibited area
(b) fly over a restricted are except in accordance with such conditions as are contained in the notice proclaiming the area."

— — —

"(4) Any person who wilfully flies over a prohibited area or who wilfully fails to comply with any notice given under this regulation commits an offence against these regulations."

NORWAY.

As a general rule, Norway claims jurisdiction over the airspace above its territory [71]. The Norwegian Law on Air Traffic (*Luftfartlov*), of December 7, 1923 [72], provides in Sec. 2:

"Aerial navigation within Norwegian territory may take place only in accordance with the provisions of the present law."

The same claim of sovereignty is explicitly expressed in the Rules for the Entry of Foreign Warships and Military Aircraft into Norwegian Territory during Peace Time [73]. Sec. 3 of these rules states:

"Norwegian territory under these rules shall be understood to be all Norwegian land and sea territory and the airspace above it."

Norway is also a party to the International Convention on Air Navigation of October 13, 1919 [74], and the Convention on International Civil Aviation of December 7, 1944 [75]. The latter provides in Art. 1:

"The contracting States recognize that every State has complete and exclusive sovereignty over the airspace above its territory."

These provisions are supplemented by Sec. 3 of the Air Traffic Act which provides that "aerial navigation within Norwegian territory may take place only with a craft which either has Norwegian nationality under Sec. 4 or Sec. 43, or meets the conditions of Sec. 15 or Sec. 16 (see also Sec. 43, last paragraph)".

Under the cited provisions only aircraft of Norwegian nationality or of such foreign nation which has an agreement with the Norwegian government are permitted to fly over Norwegian territory. Section 4 of the Air Traffic Act defines Norwegian aircraft as one which is registered with the proper Norwegian authorities and entered in the official Aircraft Register.

Section 6 of the same Act contains the requirements for aircraft registration in Norway:

"An aircraft may be registered in this realm only when it has a Norwegian owner. The following are considered Norwegians:
(a) the Norwegian state;
(b) Norwegian local government bodies;
(c) corporations and foundations, the board of which are entirely Norwegian and are domiciled in Norway;
(d) Norwegian nationals;
(e) companies of unlimited liability, unlimited and limited partnerships, associations and companies of limited liability which are not joint stock companies with limited and general shareholders or joint stock companies, if they are composed exclusively of Norwegian nationals;
(f) joint stock companies with limited and general shareholders, if the general shareholders are Norwegian nationals;
(g) joint-stock companies if the company's main office is in Norway and the members of its board are Norwegian nationals domiciled in the realm.
An aircraft may not be registered in this realm if it has acquired the nationality of another state."

Aircraft used exclusively for Norwegian government service (police, military, postal or customs aircraft) are considered possessing Norwegian nationality although they need not be entered in the Aircraft Register, (Sec. 43, Par. 1 of Air Traffic Act).

Aircraft which do not possess Norwegian nationality are allowed to fly over Norwegian territory upon the following conditions (Sec. 15):

"(a) if the craft possesses a nationality and registration certificate or a corresponding certificate from a public authority of a state with which an agreement has been concluded to the effect that a craft registered in this state shall have the right to such flight, or

"(b) if the proper Ministry, upon an application of the party for whose account the craft is operated, has given permission for the craft to make such flight."

It is furthermore provided in this section that:

"The permission mentioned under (b) shall be good for one month and may be renewed for one month at a time.

"If there is sufficient reason, the permission may be revoked at any time. It may only be granted if the craft, in regard to airworthiness, meets requirements which are equal to those imposed on Norwegian craft under the present law.

"A craft which is being used for air traffic within Norwegian territory according to subsection (a) of this section shall display nationality and registration symbols in accordance with regulations approved by the King in accordance with the agreement with the government concerned, or, in accordance with subsection (b), shall display identification marks in accordance with regulations issued by the appropriate Ministry.

"The King or a person authorized by him decides if, and, in such a case, on which conditions aircraft used to maintain aerial communications in the interest of the activities of the League of Nations, shall be admitted to Norwegian territory."

In addition, special permission for flights over Norwegian territory may be granted to aircraft used in experimental flights and tests of airworthiness, even if such an aircraft would not qualify for registration under other provisions of the law (Sec. 16).

Special rules are applicable to military aircraft of foreign states. As a general rule, such aircraft may come into Norwegian airspace or fly over Norwegian territory only if proper permission has been obtained through regular diplomatic channels [76]. Exceptions to this rule are aircraft which (a) have aboard the head of a foreign state or his official representatives, (b) accompany such an aircraft, (c) are stationed aboard a foreign warship during the latter's stay in Norwegian waters, or (d) are in distress [77].

The Norwegian Air Traffic Act lists only three kinds of aircraft; the airplane (landplanes and hydroplanes), the dirigible airship and the free balloon (Sec. 1). The Directives on Identification Marks for Civil Aircraft [78] define balloons and airships as aircraft which are lighter than air, and an airplane as a power propelled aircraft which is heavier than air and is held afloat by aerodynamic reaction upon its wings (Sec. 1). However, the statute does not specify that the airplane has to be a manned craft. Furthermore, the above-mentioned directives include under the term "aircraft" 'every apparatus which may be held afloat in the atmosphere by any reaction from the air'. This seems to include also rocket missiles within the term of "aircraft" and thus subject them to the Air Traffic Act and the claim of Norwegian sovereignty in the airspace above Norwegian land and sea territory.

Norwegian land territory is fixed by international agreements with the neighboring countries. On the sea Norway claims sovereignty over a territory extending one geographical mile (7420 m) from its coast [79]. The statutory basis for this claim is the Proclamation of February 25, 1812 [80]. For specific purposes, as for instance in regard to fishing rights, the Norwegian Government has established a demarcation line described in detail in the Royal Regulations of July 12, 1935 [81], and July 18, 1952 [82].

PAKISTAN.

Pakistan adopted the Aircraft Act. 1934 in force in India when it became a Republic in 1948. Therefore, Sec. 17 of the Act quoted in the paragraph on India applies for Pakistan.

POLAND.

The following are translations from laws dealing with "airspace" which were, or have been, in force in Poland.

I. The Convention on the Regulation of Air Navigation signed in Paris on October 13, 1919 (*Dz. U.* [Journal of Laws] of 1929, No. 6, Law No. 54):

> *Sec. 1.* The High Contracting Parties recognize that every Power has complete and exclusive sovereignty over the airspace above its territory.

The said Convention was renounced by Poland on April 5, 1948 (*Dz. U.* No. 21, Law No. 151).

II. The Order of President of the Republic of Poland of March 14, 1948, on the Aviation Law (*Dz. U.* of 1935, No. 69, Law No. 437):

> *Sec. 1.* The sovereignty over the airspace within the State boundaries including the territorial waters belongs to the Republic of Poland.

III. Decree of March 23, 1956, on the Protection of State Boundaries (*Dz. U.* No. 9, Law No. 51):

> *Sec. 1.*
> ———
> (2) The boundary line also separates vertically the airspace, the waters and the interior of the earth.

ROMANIA.

Romania enacted an Air Code in 1953 (Decree No. 516 of December 5, 1953; *Buletinul Oficial* No. 56 of December 30, 1953), which provides:

> *Art. 1.* The Romanian People's Republic has sole and full sovereignty over its airspace.

Thus the Romanian People's Republic has adopted the theory of complete sovereignty over the airspace above its territory.

Survey of Legal Opinion on Extraterrestrial Jurisdiction

SAUDI ARABIA.

Saudi Arabia is among the few countries which have no basic aviation law at all [83]. Saudi Arabia has also not adhered to the Convention on International Civil Aviation signed in Chicago on December 7, 1944, and has not even participated in the 1954 meetings of the Sub-Committee for Civil Aviation Matters of the Permanent Communications Commission of the League of the Arab States [84].

UNION OF SOUTH AFRICA.

The Aviation Act, No. 16 of 1923 as amended by Act No. 41 of 1946 and No. 42 of 1947 in Union Statutes 1910—1947, contains two sections regulating flight over its territory. Article 3(1) provides:

"The Governor General may make regulations relating to all or any of the following matters or things, namely . . .

"(1) measures for preventing aircraft flying over prohibited areas or entering or leaving the Union in contravention of any provision of this Act.

―― ― ―

"5(1) In time of war, whether imminent or actual, or within six months after a state of war ceased to exist, or in time of great national emergency, the Governor General may:

"(a) by proclamation in the *Gazette,* declare that the Union or any parties of the Union including the territorial waters thereof, shall be restricted area for the purpose of this section

"(b) issue orders and instructions in respect of any restricted area or part thereof

"(i) regulating, restricting or prohibiting the navigation of all or any description of aircraft."

SWEDEN.

Swedish law in regard to the sovereignty over the airspace is based on the presumption of an unrestricted sovereignty of the state over the airspace above its territory. This territory includes both the land and the territorial waters [85]. The basic Swedish aviation act, the Royal Ordinance on Aviation (*Förordning om luftfart*) of May 26, 1922 [86], provides in Sec. 3:

"Air traffic over Swedish territory may be undertaken only by an aircraft which either is of Swedish nationality according to Sections 4 or 38 or meets the conditions of Sections 14 or 15."

According to these provisions in Sections 14 or 15, only the following aircraft are allowed over the Swedish territory:

(1) an aircraft of Swedish nationality;
(2) an aircraft possessing a nationality and registry certificate of a foreign state, provided there is an agreement permitting the aircraft of this foreign state to fly over Swedish territory, between Sweden and the foreign state concerned;
(3) an aircraft in possession of a special permit from the Air Traffic Office (*Luftfartstyrelsen*);
(4) an aircraft in possession of a special permit from the Air Traffic Office for the purpose of testing the airworthiness or other technical conditions of the aircraft over Swedish territory.

In the same vein, Sec. 1 of the Law of September 7, 1914, concerning the Prohibition of Air Traffic over Swedish Territory (as amended by Law of October 14, 1939), states:

"The King shall be authorized, if it proves to be necessary, to order that no flying over Swedish territory by aircraft other than those belonging to the Swedish State or operated for it may take place without a permit from the King or from an authority designated by the King" [87].

Swedish legislation on aviation and aircrafts does not contain any definitions concerning the extent of the airspace included within the claim of sovereignty. The Royal Ordinance of 1922 includes under the term "aircraft" *(luftfartyg)*, airplanes (landplanes and seaplanes), mechanically driven balloons *(motorballonger)* and free balloons *(fria ballonger)* (Sec. 1). Regulations issued for the implementation of this ordinance [88] give a more exact definition of aircraft coming within the meaning of the 1922 ordinance. Sec. 1 of the Regulation of 1928 explains, for instance, that this term includes also airships and captive balloons. Since these regulations and laws do not require that the "aircraft" be manned, it has been alleged by some legal writers in the field that the term "aircraft" in the above ordinance and regulation includes all radio-controlled unmanned airplanes and rocket ships [89].

Such an interpretation of the term "aircraft" serves to emphasize the point that in Swedish law there is set no limit to the sovereignty of the state in "airspace" *(luftrummet)*. No indications, however, could be found in the statutory law or legal writings of whether or not the claim of state sovereignty is considered to extend beyond the earth's atmosphere into the so-called "outer space " [90].

A new Aviation Law *(Luftfartslag)* was enacted in Sweden on June 6, 1957 [91]. It repeals and replaces the earlier laws on this subject, among them the Royal Ordinance on Aviation *(Förordning om luftfart)* of May 26, 1922, with its later amendments, and the Law of September 7, 1914, relating to the prohibition of Air Traffic over Swedish Territory.

The Aviation Law of 1957 authorizes the King to set the effective date of the law. No information is available at the present time that such a date has been set by any Royal enactment [92]. No regulations to implement the provisions of the new law have been yet issued.

The new law does not effect any changes in the Swedish attitude toward its claim of sovereignty over the airspace above its territory as described in the earlier report. Mainly, it codifies and modernizes the existing law in regard to aviation. The pertinent sections of the new law are given below in translation.

AVIATION LAW *(Luftfartslag)* June 6, 1957
(Translation from *Svensk Författningssamling* 1957, No. 297)
FIRST PART: CIVIL AVIATION

Chapter 1: Introductory Provisions
Sec. 1. Aviation here in the state is allowed only under restrictions and conditions which follow from the present Law and from provisions published in its implementation or which otherwise are provided for in statutes and regulations.

Sec. 2. No aircraft other than those of Swedish nationality or of a nationality of a foreign state with which an agreement has been concluded in regard to the right to aviation within Swedish territory shall be allowed to fly.

If special reasons exist, the King or, under his authorization, the Aviation Board shall be entitled to allow aviation without regard to the provisions of the first paragraph and also to establish conditions therefor.

Sec. 3. For military reasons or for reasons of public order and security, the King shall be entitled to restrict or prohibit flying within a certain part of the country. In accordance with an authorization from the King, the Aviation Board may also, for reasons of public order and security, restrict or for not more than two weeks forbid flying within a certain territory.

Under extraordinary circumstances or if it is necessary for reasons of public security, the King shall be authorized to restrict or forbid flying temporarily·within the whole country.

Chapter 2: Registration of Aircraft, their Nationality and Identification Registration
Sec. 2. An aircraft may be registered here in the state only if its owner is the Swedish state, a Swedish local government unit or other similar body, a Swedish national or Swedish estate of a deceased, or a company, association, other society, foundation or other similar institution which possesses Swedish nationality.

In regard to an aircraft which as a rule has its point of departure within the country (Sweden), the King shall be authorized to allow its entry in the register regardless of the provisions of the first paragraph.

Sec. 3. An aircraft which is registered in a foreign state may not be registered in Sweden unless it has been removed from the register of the foreign state.

— — —

Sec. 12. After an aircraft has been entered in the Aviation Register it shall possess Swedish nationality.

The Aviation Board shall issue a certificate evidencing the entry in the Aviation Register (certificate of nationality and registration).

PART TWO: MILITARY AVIATION AND AVIATION FOR OTHER GOVERNMENT PURPOSES

Chapter 15

Sec. 1. With regard to matters covered by the respective civil aviation regulations in the present Law, the King or a delegate of the King shall publish rules in regard to aviation by military aircraft and also in regard to military airfields and other military ground organization.

— — —

Sec. 3. The King or his delegate shall establish whether and under what conditions foreign military aircraft or other foreign aircraft which are exclusively used for governmental purposes and not for business may obtain admission to Swedish territory.

SYRIA.

Syria has adopted the theory of complete and exclusive sovereignty over the airspace above its territory, *i.e.,* over its land areas and the territorial waters adjacent thereto.

Already the Law Regulating Air Transportation and Navigation of November 28, 1948 [93] had expressed this principle. This law has been repealed and replaced by the Law Regulating Air Transportation and Navigation, 1949 [94] which reads in part [95]:

Sec. 1. The State has complete and exclusive sovereignty over the airspace above its territory. The territory of the State includes, as far as the application of this law is concerned, the territorial waters.

Sec. 2. No airplane is permitted to fly over the territory of Syria or to land unless it has been authorized to do so in conformance with the provisions of the present law.

Secs. 3, 4 and 5 of the above-mentioned law contain definitions of the term "airplane" and of government and civil airplanes.

TURKEY.

The situation with respect to the civil aviation laws of Turkey has been summed up as follows in a recent survey of national legislation on aviation since the Chicago Convention [96]:

The only source of Turkish air law is a "Regulation of Air Navigation" of September 9, 1925. Soon after the ratification of the Chicago Convention, the Ministry of Communications established a Commission entrusted with the preparation of a draft law on aviation. The draft so prepared was quite complete. Some time later, an ICAO Technical Assistance Mission was sent to Turkey, with a legal expert. That Mission does not seem to have been successful in the preparation of a basic law. Recently, a new draft conceived in general terms has been under study and will be submitted to Parliament.

The Air Navigation Regulation of September 9, 1925, reads in part:

Art. 1. The air space of the Turkish Republic is limited by its territorial frontiers and by that of the seas which constitute its territorial waters.

Art. 2. Provides that airplanes and balloons, other than Turkish, are forbidden to fly over or to land in the prohibited zones established within the Turkish air space frontiers.

Art. 3. Provides for freedom of passage within the airspace of Turkey to non-military airplanes and balloons of the States which signed the International Convention on Aerial Navigation of October 13, 1919. This convention is now obsolete and has been replaced since April 4, 1947, by the Convention of International Civil Aviation of December 7, 1944.

The Regulation of Air Navigation further contains provisions on Rules concerning Military and non-Military Aircraft (Art. 6—11), as well as specific rules on other air navigation questions.

An extensive treatise, "Turkish Commercial Aviation" by Mieczyslaw Budek has been published in 1956 [97]. It stresses the fact that Turkey has concluded twenty bilateral agreements on air Transport services with other countries [98].

By the passing of Law 4749 of June 12, 1945, the Convention on International Civil Aviation became Turkish national law [99]. Thus, Turkey has incorporated into its law also the principle of complete and exclusive sovereignty over the airspace above the territory, as specified in Article 1 of the Convention [100].

UNITED KINGDOM.

Jurisdiction over airspace of the United Kingdom is governed by common law, statute and provisions of international conventions ratified by the United Kingdom.

The common law rule is stated by Coke *On Littleton,* 4a, in these words:

> "And lastly, the earth hath in law a great extent upwards, not only of water, as hath been said, but of ayre and all other things even up to heaven; for *cujus est solum, ejus est usque ad caelum* (whoever owns the soil owns all that lies above it)."

Blackstone (*Commentaries,* volume 2, chapter 2) on the same subject says:

> "Land hath also, in its legal signification, an indefinite extent upwards, as well as downwards . . ."

British statute law relating to the control of airspace is found in the Civil Aviation Act, 1949 (12 & 13 Geo. 6, c. 67):

> "40(1) No action shall lie in respect of trespass or in respect of nuisance by reason of the flight of an aircraft over any property at a height above the ground which having regard to wind, weather and all the circumstances of the case is reasonable, or the ordinary incidents of such flight so long as the provisions of Part II [101] — and this Part of this Act and any Order in Council or order made under Part II or this Part of this Act are duly complied with.
>
> "(2) Where material loss or damage is caused to any person or property on land or water by, or by a person in, or an article or person falling from, an aircraft while in flight, taking off or landing, then unless the loss or damage was caused or contributed to by the negligence of the person by whom it was suffered, damages in respect of the loss or damage shall be recoverable without proof of negligence or intention or other cause of action, as if the loss or damage had been caused by the wilful act, neglect, or default of the owner of the aircraft.
>
> "61(1) Neither this Part of this Act nor any enactment to which this Part of this Act applied shall apply to aircraft belonging to or exclusively employed in the service of His Majesty.
>
> "Provided that His Majesty may, by Order in Council, apply to any such aircraft, with or without modification, any of the said enactments or any Order in

Council, apply to any such aircraft, with or without modification, any of the said enactments or any Orders in Council, orders or regulations made thereunder."

International Law. According to Halsbury's *Laws of England* (3rd edition, London, Butterworth & Company, 1953), *6, 5:* "A large part of the English law relating to civil aviation is directly or indirectly derived from the provisions of conventions upon international aerial navigation and cognate matters. These conventions are the outcome of agreements between numbers of sovereign states providing for the mutual and uniform regulation of air traffic and matters incidental thereto. They are, however, in the nature of multilateral treaties between states and form no part of English municipal law save insofar as they are incorporated therein by domestic legislation."

Sovereignty over airspace is established for all nations which have ratified the Convention on International Civil Aviation known as the Chicago Convention. This Convention was ratified by the United Kingdom and came into force on April 4, 1947.

> *Art. 1.* The contracting States recognize that every State has complete and exclusive sovereignty over the airspace above its territory.
>
> *Art. 2.* For the purposes of this Convention the territory of a State shall be deemed to be the land areas and territorial waters adjacent thereto under the sovereignty, suzerainty, protection or mandate of such State.

In conclusion, the opinions of Sir Percy Winfield and Sir Arnold D. McNair on the subject of airspace under British law are given. *Winfield on Tort* (6th edition, London, Sweet & Maxwell, 1954), with respect to trespass by aircraft of airspace under British law, states (page 380):

> "As to common law, there is not a single reported English decision (as distinct from dicta) discoverable."

Sir Arnold D. McNair in the *Law of the Air* (2nd edition, London, Stevens and Sons, 1953) on page 31, sums up the matter in these words:

> "But can space—whatever space may be—become the subject of ownership? The point must be viewed with doubt. Certainly the 'ownable' contents of space may be owned, whether they are minerals below the surface of the earth or buildings above it. It cannot be definitely affirmed that the common law is committed to the view that mere abstract space can be the subject of ownership apart from its contents. . . .
>
> "It is suggested that we must reject the theory of the ownership of the whole column of air space to an indefinite height by the owner of the surface (including in the term 'surface' the top floor of any structure erected upon it).
>
> "It is suggested further that there are only two theories which can be accepted without doing violence to that common sense for which the common law is famous. Those two theories are: (1) that *prima facie* a surface-owner has ownership of the fixed contents of the air space and the exclusive right of filling the air space with contents, and alternatively, (2) the same as (1) with the addition of ownership of the air space within the limits of an 'area of ordinary user' surrounding and attendant upon the surface and any erections upon it."

UNITED STATES.

The Air Commerce Act of 1926, as amended, contains the following declaration of sovereignty of airspace:

> US Code 49:176. *Foreign aircraft*
> (a) *Sovereignty of airspace declared; navigation of foreign military aircraft.*
> The United States of America is declared to possess and exercise complete and exclusive national sovereignty in the air space above the United States, including

the air space above all inland waters and the air space above those portions of the adjacent marginal high seas, bays, and lakes, over which by international law or treaty or convention the United States exercises national jurisdiction. Aircraft a part of the armed forces of any foreign nation shall not be navigated in the United States, including the Canal Zone, except in accordance with an authorization granted by the Secretary of State.

(b) *Navigation of foreign civil aircraft; reciprocity; engagement in air commerce.*
Foreign aircraft, which are not a part of the armed forces of a foreign nation, may be navigated in the United States by airmen holding certificates or licenses issued or rendered valid by the United States or by the nation in which the aircraft is registered if such foreign nation grants a similar privilege with respect to aircraft of the United States and only if such navigation is authorized by permit, order, or regulation issued by the Civil Aeronautics Board hereunder, and in accordance with the terms, conditions, and limitations thereof. The Civil Aeronautics Board shall issue such permits, orders, or regulations to such extent only as the Board shall find such action to be in the interest of the public: *Provided, however,* That in exercising its powers hereunder, the Board shall do so consistently with any treaty, convention or agreement which may be in force between the United States and any foreign country or countries. Foreign civil aircraft permitted to navigate in the United States under this subsection may be authorized by the Board to engage in air commerce within the United States except that they shall not take on at any point within the United States, persons, property, or mail carried for compensation or hire and destined for another point within the United States. Nothing contained in this subsection shall be deemed to limit, modify, or amend section 482 of this title but any foreign air carrier holding a permit under section 482 of this title shall not be required to obtain additional authorization under this subsection with respect to any operation authorized by said permit. (May 20, 1926, ch. 344, §6, 44 Stat. 572; June 23, 1938, ch. 601 §1107 (i) (1), (3, 4, 5), 52 Stat. 1028; as amended Aug. 8, 1953, ch. 379, 67 Stat. 489.)

By an Act approved July 9, 1937, (50 Stat. 486) Congress amended the Canal Zone Code by adding the following section to Title 2:

T. 2-§14. *Air Navigation.* The Government of the United States is hereby declared to possess, to the exclusion of all foreign nations, sovereign rights, power, and authority over the air space above the lands and waters of the Canal Zone. Until Congress shall otherwise provide, the President is authorized to make rules and regulations and to alter and amend the same from time to time governing aircraft, air navigation, air-navigation facilities, and aeronautical activities within the Canal Zone. Any person who shall violate any of the rules or regulations issued in pursuance of the authority contained in this section shall be punishable by a fine of not more than $500, or by imprisonment in jail for not more than one year, or by both. (As added July 9, 1937, ch. 470, sec. 1, 50 Stat. 486 [US Code, title 48, sec. 1314a].)

Sections entitled "Security Provisions" were added to the Civil Aeronautics Act by a law enacted September 9, 1950 (64 Stat. 825) in which section 1203 provides for security control of air traffic:

US Code 49:703. *Security control of air traffic.* The Secretary of Commerce is authorized to establish such zones or areas in the airspace above the United States, its Territories, and possessions (including areas of land or water administered by the United States under international agreement) as he may find necessary in the interests of national security; and may, after consultation with the Department of Defense and the Board, by rule, regulation, or order within such zones or areas, prohibit or restrict flights of aircraft which he cannot effectively identify, locate, and control with available facilities: *Provided,* That the Secretary of Commerce shall consult with the Department of State before exercising the authority provided in this section with respect to areas of land or water administered by the United States under international agreement. (June 23, 1938, ch. 601, title XII, § 1203, as added Sept. 9, 1950, ch. 938, 64 Stat. 825.)

Sec. 16 of the Federal Airport Act of May 13, 1946, provides for the use of Government-owned lands including interest in airspace:

> US Code 49:1115. *Requests for use of Government-owned lands; determination; execution of conveyance.*
>
> (a) Whenever the Administrator determines that use use of any lands owned or controlled by the United States is reasonably necessary for carrying out a project under this chapter, or for the operation of any public airport, he shall file with the head of the department or agency having control of such lands a request that such property interest therein as he may deem necessary be conveyed to the public agency sponsoring the project in question or owning or controlling the airport. Such property interest may consist of the title to or any other interest in land or any easement through or other interest in air space.
>
> (b) Upon receipt of a request from the Administrator under this section, the head of the department or agency having control of the lands in question shall determine whether the requested conveyance is inconsistent with the needs of the department or agency, and shall notify the Administrator of his determination within a period of four months after receipt of the Administrator's request. If such department or agency head determines that the requested conveyance is not inconsistent with the needs of that department or agency, such department or agency head is authorized and directed, with the approval of the President and the Attorney General of the United States, and without any expense to the United States, to perform any acts and to execue any instruments necessary to make the conveyance requested; but each such conveyance shall be made on the condition that the property interest conveyed shall automatically revert to the United States in the event that the lands in question are not developed, or cease to bo used, for airport purposes. (May 13, 1946, ch. 251, §16, 60 Stat. 179.)

I. Preliminary Remarks

USSR.

The definition of the airspace of the Soviet Union and its sovereignty is given in the first Soviet Air Code of April 27, 1932, (Secs. 1—2; *see infra* II) and restated in the Air Code of September 3/13, 1935, which is mentioned as the law still in force in the latest available university text book on international law, printed in 1935 (Lisovskii, V. I., *Mezhdunarodnoe pravo*, Kiev, 1955, pp. 158—159).

The Code of 1935 contains also provisions concerning international flights over the airspace of the USSR which are translated under III. These provisions are essentially similar to those of the Code of 1932.

Furthermore, the above-mentioned text book on international law elaborated also on the concepts stated in the Code of 1935, and related passages are translated under IV.

Special rules regulate the entry and flight in the airspace of the USSR of military aircraft. Their text is not available but their gist is given in a university text book on international law of 1947, and the rules are mentioned in a similar textbook of 1951. The corresponding passage from the textbook of 1947 is translated *infra* under V.

II. Air Code of the USSR of April 27, 1932 (USSR Laws, 1932, No. 32, Text 194b)

> *Sec. 1.* The Air Code shall be in force within the boundaries of the land and water territory, the coastal sea water and the airspace of the USSR.
>
> The airspace of the USSR shall be understood to be the airspace over the land and water territory of the USSR and over the zone of the coastal sea waters as determined by the law of the USSR.
>
> *Sec. 2.* To the USSR shall belong the complete and exclusive sovereignty over the air space of the USSR.

III. Air Code of the USSR of September 3/13, 1935 (USSR Laws 1935, No. 34, Text 359a)

Chapter I. General Provisions

Sec. 1. To the USSR shall belong complete and exclusive sovereignty over the airspace of the USSR.

The airspace of the USSR shall be understood to be the airspace above the land and water territory of the USSR and above the zone of sea coastal waters as determined by the laws of the USSR.

Sec. 2. The Air Code of the USSR shall be in force within the boundaries of the land and water territory of the USSR, in the zone of the coastal sea waters determined by the laws of the USSR, and in the airspace of the USSR.

Chapter II. Civil Aircraft

Sec. 7. With civil aircraft shall be classed all aircraft designed for air transportation (those lighter as well as heavier than air) with the exception of the aircraft incorporated into the armed forces.

Chapter VII. International Flights

Sec. 54. Any flight of a civil aircraft during which the aircraft crosses the state boundaries of the USSR shall be considered an international flight.

To international flight shall apply general rules of flight in the airspace of the USSR with supplements and amendments stated in the present chapter.

Sec. 55. Civil aircraft not entered in the Register of the USSR may perform flights to the confines of the USSR, into the airspace of the USSR, and from the USSR beyond its boundaries only with a special permission of the Main Administration for Civil Aviation.

The following shall be indicated in the permit:

(a) Air gates through which the craft must cross the state boundary of the USSR while flying in or out as well as the height of the flight while crossing the frontier;

(b) Route by which the craft must fly as well as places for mandatory and permissible landings;

(c) Period of time of validity of the permit.

Sec. 56. Civil aircraft entered in the Register of the USSR may perform flights beyond the confines of the USSR only with the permission of the Main Administration for Civil Aviation.

Sec. 57. While performing international flights all the civil aircraft shall be guided in the customs operations by the Customs Code of the USSR and regulations issued in its implementation.

Sec. 58. To the persons arriving in the USSR and those departing from the USSR by civil aircraft shall apply the general rules for entry into the USSR, departure from the USSR, and transit through the USSR.

Sec. 59. The acceptance and release, from the point of view of custom-house and passport control, of civil aircraft which perform international flights shall be carried out at airports and seaplane ports or ports for dirigibles determined by the Main Administration for Civil Aviation in agreement with the People's Commissariat for Foreign Trade and People's Commissariat of the Interior of the USSR.

(*Note:* All the People's Commissariats were renamed Ministries in 1946.)

Sec. 60. If a civil aircraft, because of an elemental emergency or for any other reason, crosses the state frontier of the USSR outside the air gates indicated or not at the indicated height or appears to be outside of the air route established for it, such aircraft should immediately upon ascertaining this fact or upon receipt of a signal requiring its grounding, give a signal of distress, lower its flight, and ground at the necessary place suitable for it.

Civil aircraft which is grounded under the conditions stated in the present section may continue its flight only with the permission of the Main Administration for Civil Aviation or its corresponding territorial agencies.

If the aircraft does not obey the signal requiring its grounding then, after a second signal, it will be forced to be grounded.

Sec. 61. If a civil aircraft which is making an international flight suffers a plane wreck or makes a forced landing before the mandatory landing at the place determined for custom-house and passport control of the given craft, then in the absence at the place of landing of a custom-house agency, the local authorities must take all the necessary steps for the protection of the cargo, luggage and the other property on board the craft and for the fulfillment of all the required passport formalities.

The same rule shall apply in instances where an aircraft making an international flight suffers a plane wreck or makes a forced landing in the USSR after fulfillment of the custom-house and passport formalities required for exit.

The Main Administration for Civil Aviation shall issue, in agreement with the People's Commissariat of Foreign Trade and People's Commissariat for the Interior of the USSR, Rules in implementation of the present section.

(Note: All the People's Commissariats were renamed Ministries in 1946.)

Sec. 62. Civil aircraft not entered into the Register of Civil Aircraft of the USSR may be subject to examination for the determination of its technical fitness for flights:

(a) If the USSR has no agreement with the State concerned regarding recognition in the USSR of the documents issued by that State on technical fitness of the civil aircraft for flights;

(b) In case of a plane wreck and disclosure of technical defects of the aircraft.

Sec. 63. Regular international flights along the international air routes shall be conducted according to the rules established in treaties made by the government of the USSR with foreign States and organizations and on problems not covered by such treaties according to the rules of the present Code.

Sec. 64. The Main Administration for Civil Aviation shall have the right to establish for foreign civil aircraft which are making single international flights of a special character departures from the rules of the present Code.

Sec. 65. The Main Administration for Civil Aviation shall establish the rules concerning international flights which apply to the free aerostat.

Sec. 66. Laws and rules in force in the USSR shall apply to the foreign civil aircraft, their crews and passengers while they are flying in the airspace of the USSR.

Sec. 67. Those guilty of violation of rules of international flights (flights into the USSR and flying out of the USSR without permission, failure to follow the air route indicated in the permit, the place of landing, the air gates, the height of flights, and the like) shall be prosecuted in criminal court under the laws of the USSR or punished in administrative action under section 95 of the present Code.

Chapter IX. Shipping by air of Passengers, Luggage, Cargo and Mail

Sec. 76. To the international shipping of passengers, luggage and cargo shall apply the rules of international conventions made by the USSR and, in the USSR and, in the absence of such rules, the provisions of the present Code and the rules issued in its implementation.

Chapter X. Fines Imposed by Administrative Action by the Agencies of Civil Aviation

Sec. 95. For violation of rules of international flights in cases where there is no reason for prosecution the violators before a criminal court, the chief of the Main Administration for Civil Aviation shall have the right to impose upon the violator in administrative action a fine up to 3000 rubles.

IV. Latest University Textbook on International Law by Lisovskii, V. I., *Mezhdunarodnoe pravo,* Kiev, 1955, p. 159:

"The air sovereignty means essentially the right of the State to prohibit or restrict the flights of foreign aircraft; its right to defend and protect the airspace; [and] the right for a monopoly of flights over its territory.

"The experience of the military air operations during the first, and especially the second World War supports the thesis that air as well as the territory should be the objective of the military defense of the country. Practical application by one or another State of the principle of the freedom of the air would mean extending to its future adversaries the possibility of complete control (in technical and military respects) of the airspace over the territory and coastal waters of the country.

"The sovereignty of a State extends to the stratosphere (that is, the stratum of air from 11 to 75 kilometers from sea level) which is over the territory of the State as well as to the troposphere (that is, stratum of air up to 11 kilometers from the sea level) over this territory because at the present level of technique of aviation not only peaceful flights may be performed in the stratosphere, but also military operations. Consequently, the State must have the right to regulate the traffic of the foreign vessels also in this stratum of the air."

V. Durdenevskii, V. N., and Krytov, S. B., *Mezhdunarodnoe pravo,* Moskva, 1947, pp. 273—274:

"10. *Flights of Foreign Military Aircraft.* The procedure and conditions of admittance of flights of foreign military aircraft are determined by the provisional rules issued by the People's Commissariat for Defense May 1, 1934. According to these rules a foreign military aircraft, for flight in the airspace of the USSR, must obtain in advance permission from the Soviet Government which should be asked through diplomatic channels. Military aircraft which arrived according to this procedure enjoy *extraterritoriality.* The flight is permitted through air gates and in accordance with the air route indicated in the permit. During the flight the rules prohibit the sending of coded messages, having on board munitions and other means of attack, ejecting by parachute people and property, sketching maps and making moving pictures. During war an exception may be made for foreign allied aircraft."

YUGOSLAVIA.

I. Statutory Provisions

1. Yugoslavia first enacted a law on aviation on February 22, 1928 [102]. According to this law foreign aircraft could fly over the territory of Yugoslavia and its territorial waters provided such right was recognized by international agreements or a special permit was granted (Sec. 44). However, the terms "airspace" and "atmosphere" did not appear in the statute, and no definition of them was to be found in later provisions.

The provisions of the Laws of 1928 insofar as they are not in contradiction with the new Communist legislation, are still in force [103].

2. The present Communist Government issued a decree on civil aviation in 1946 [104]. The decree states that aircraft of foreign countries may fly through the airspace of the FPRY provided they follow the fixed corridors (Sec. 6).

3. The law on territorial waters was passed in 1948 [105]. It provides that aircraft of foreign countries may fly above the territorial waters of the FPRY only if they comply with the existing provisions applicable to flights over its territory (Sec. 11). The territorial waters of the FPRY extend seaward six miles from the watermark of the seacoast, the islands, the outermost points of the port installations, and the line of the inland waters (Sec. 5).

II. International Agreements Adopted by Yugoslavia

1. The first significant pronouncement on the law of aviation came after World War I when a convention for the regulation of aerial navigation was adopted in Paris on October 13, 1919. This convention was ratified by Yugoslavia in 1926 [106]. It remained the basic international law on the subject of aerial navigation until it was superseded by the International Civil Aviation Convention. Questions of sovereignty in the air were passed upon as well as the rights of nationals of one country to fly in or through the airspace of another country. Thus, it provided that every Power has complete and exclusive sovereignty over the airspace above its territory (Art. 1).

In connection with the principle of sovereignty set forth in Article 1, the convention stated that each contracting state is entitled to prohibit the aircraft of the other contracting states from flying over certain areas of its territory for military reasons or in the interests of public safety. Although the prohibition was limited only "for military reasons" and "in the interests of public safety", according to an authority these terms, having a broad meaning, might be applied by the contracting states at their discretion [107].

2. The International Civil Aviation Convention in Chicago, December 7, 1944, superseded the Convention of Paris of 1919. Yugoslavia ratified this convention in 1953 [108].

The principal innovations resulting from this new agreement are: (a) the creation of a world-wide aviation organization with power to function in the fields of safety and economic matters; (b) the opening up of the air for transit purposes; (c) the opening up of the air for international air transportation purposes.

The convention provided that every state has complete and exclusive sovereignty over the airspace above its territory, which includes the land areas and territorial waters adjacent thereto under its sovereignty (Art. 8, Secs. 1 and 2).

III. Opinions of Legal Writers

According to Dr. Milan Bartôs, an authority on international law, the state has an uncontested right of sovereignty over the airspace which extends vertically upward within the boundaries of its territory and territorial waters. The height of that space is irrelevant; it includes the sphere and the stratosphere [109].

Dr. Bertold Eisner, another authority on conflict of laws, stated that it is generally accepted that sovereignty over the space above its territory belongs to a state. Therefore, the airspace may be considerd as a part of the state territory [110].

References

[1] This authority is now vested in the Minister of Transport and Communications; Burma being an independent sovereign nation since January 1948.

[2] *Z. Skalsky* and *D. Fedor* in Právnik, 1956, p. 711.

[3] *Ibid.*, p. 713.

[4] No. 356 of July 25, 1951 (*Anordning om fremmede krigsskibes og militaere luftfartøjers adgang til dansk område under fredsforhold*), in *Love og Anordninger,* Collection of laws and regulations, Part A, 1951, p. 1842.

[5] See *Poul Andersen*. Dansk Statsforfatningsret (Danish Constitutional Law), I. 2nd ed. København, 1949, p. 106.

[6] *Lov No. 175 om Luftfart,* in the version of the Proclamation No. 251 of August 6, 1937, *Danmarks Love* (Laws of Denmark), 1665—1949. København, 1950, pp. 1565—1573.

[7] As an exception, custom, police and postal aircraft exclusively in the service of the government and military aircraft need not be entered in the Aircraft Register, but nevertheless possess Danish nationality (Sec. 42 of the Aviation Act).

[8] Sec. 1. Also Sec. 1 of the Regulation No. 490 under the Aviation Act (*Anordningen om Luftfart*) of September 11, 1920, *Love og Anordninger,* Part A, 1920, p. 1236. Sec. 39 amended by Regulation No. 349 of September 6, 1947.

[9] *Samling af Bekendtgørelser* (Collection of Regulations) vol. 5 (1956), pp. 493—494.

[10] Sec. 13 of the Decree concerning the Entry of Foreign Warships and Military Aircraft.

[11] Ratified March 30, 1948, *Lovtidende C,* part of the Danish official gazette containing the international treaties and agreements, 1948, text No. 17. Denmark adhered also to the Paris Convention of 1919. See also Andersen, *op. cit.,* p. 116.

[12] *Ibid.,* pp. 106—110.

[13] *Ibid.,* pp. 11 ff. Also *Roald Johansen,* "*Rettshåndhevelsen på sjøterritoriet*", (Law Enforcement in Territorial Waters), *Nordisk Administrativt Tidskrift* vol. 38 (1957), no. 2, pp. 100 ff. and *Thorsten Kalijarvi,* "Scandinavian Claims to Jurisdiction Over Territorial Waters", American Journal of International Law, vol. 26 (1932) p. 61. Also law no. 500 of December 19, 1951 on Salt Water Fishing, *Lovtidende A,* 1951, No. 500.

[14] Law No. 277 of May 27, 1950, concerning the Exercise of Trades in Greenland *(Lov om udøvelse af erhverv i Grønland), Love og Anordninger*, Part A, 1950, p. 1134, and the Regulation No. 292 of Novemver 11, 1953, *ibid.*, 1953, p. 2168. See also *Johansen, op. cit.*, p. 101.

[15] *Andersen, op. cit.*, p. 111; *Johansen, op. cit.*, p. 101.

[16] La République Egyptienne. Ministère de la Guerre. Administration de l'Aviation Civile. *Recueil des Legislations Egyptiennes de l'Aviation. Lois. Décrets-Lois, Décrets, Arrêtés Ministériels. Accords Bilatéraux. Conventions Internationales.* Le Caire, Imprimerie Nationale, 1954, p. 14. This is an extensive (929 pages) collection of Egyptian legislation on aviation compiled by the Administration of Civil Aviation of the Egyptian Ministry of War. It includes not only the laws, decrees having the force of laws, decrees and departmental decisions, but also all bilateral treaties and international conventions to which Egypt is a contracting party.

[17] Ratification documents deposited at the US Government Archives March 13, 1947, No. 15 *(Journal Officiel* 20). The Egyptian Government approved the Convention by Law of March 2, 1947 and has promulgated the convention by Decree of May 26, 1947, making April 12, 1947 the day when the Convention became effective *(Journal Officiel* 48 and 57), see *op. cit. supra* n. 1, p. 513, footnote 1, para. 2.

[18] *Eugène Pépin*, Development of the National Legislation on Aviation Since the Chicago Convention, in 24 The Journal of Air Law and Commerce (No. 1, Winter 1957), p. 16.

[19] *Nathan Marein.* The Ethiopian Empire Federation and Laws, Rotterdam, Royal Netherlands Printing and Lithographing Company Late J. Vürtheim & Son Ltd., 1955, p. 456.

[20] *Ibid.*, pp. 319—320. The Charter of the Imperial Ethiopian Aero Club appeared in the *Negarit Gazeta* dated June 28, 1950 under General Notice No. 133/50, *ibid.*, p. 53, under (10).

[21] The Charter of the Ethiopian Airlines was published in the *Negarit Gazeta* dated December 30, 1945 under General Notice No. 59 of 1945 — *Ibid.*, p. 53, under (3).

[22] Ratified by Finland on April 22, 1949, *Suomen Asetuskokoelma*, Finland's law gazette, 1949, item No. 331 and *Suomen Asetuskokoelman Sopimussarja*, Finnish Treaty Collection, 1949, item No. 11. Finland had joined also the Paris Convention of October 13, 1919. See *ibid.*, 1931, item No. 18 and 1936, item No. 11.

[23] See *E. J. Manner,* "*Eräitä ilmatilan oikeudellista asemaa ja käyttämistä koskevia näkökohtia*" (Some views regarding the Legal Status and Use of Airspace), *Lakimies* (Helsinki), Vol. 46 (1948), p. 657.

[24] Text in *Erkki Ailio,* (editor), *Suomen Laki* (Finnish Law), Vol. I. Helsinki, 1955, pp. 416—417. In general on aviation legislation *see Rudolf Beckman,* "*Ilmalainsäädäntö*" (Aviation Legislation), *Lakimies* Vol. 46 (1948), pp. 548—601.

[25] *Asetus ulkomaisten sota-, kauppa- ja ilmalusten käynnistä rauhan aikana;* text in *Suomen Laki,* Vol. I, pp. 437—440. Amended by regulation of August 18, 1956, *Suomen Asetuskokoelma* 1956, item No. 466.

[26] Text in *Kauko Sipponen,* (editor), *Suomen Laki,* Vol. II. Helsinki, 1956, pp. 1725—1726.

[27] *Asetus Suomen aluevesien rajoista annetun lain soveltamisesta,* ibid., pp. 1726—1727.

[28] *Bundesgesetzblatt* (Collection of Laws) abbreviated BGBl. 1955, Part II, p. 405.

[29] According to the index to the laws presently in force in Germany, by F. Schlegelberger, *Das Recht der Gegenwart,* Berlin, 1957.

[30] *Reichsgesetzblatt* (Collection of Laws) abbreviated RGBl. 1936, Part I, p. 653.

[31] RGBl. 1938, Part I, p. 1246.

[32] RGBl. 1943, Part I, p. 69. None of these amendments have any bearing on the question under investigation.

[33] RGBl. 1936, Part I, p. 659.

[34] RGBl. 1937, Part I, p. 432, 815, 1387 respectively.

[35] RGBl. 1938, Part I, p. 1237.

[36] RGBl. 1954, Part I, p. 302 rectified p. 372.

[37] RGBl. 1955, Part I, p. 321.

[38] Translated from: *Deutsche Luftfahrtgesetzgebung*, by *A. Wegerdt*. (2nd ed. Munich, Pohl & Co., [1955].)

[39] As amended, *See supra*.

[40] Translation taken from The German Civil Code, by *Chung Hui Wang*, London, 1907, p. 202.

[41] As amended, *See supra*.

[42] Constitutions of Nations, by *Amos J. Peaslee*, 2nd ed. Vol. II (The Hague, Nijhoff, 1957) p. 42.

[43] "The Development and Present State of German Air Law" in The Journal of Air Law and Commerce, Vol. 23, No. 2, p. 188 *et seq*.

[44] Announcement of ratification of April 2, 1947, published in *Stjornartidindi*, Iceland's official gazette, 1947, Part A, p. 150, under No. 45.

[45] *Ibid.*, 1929, p. 81, No. 32. Amendments were enacted on May 28, 1941 (*ibid.*, 1941, p. 21, No. 20) and on May 17, 1945 (*ibid.*, 1947, p. 184, No. 49.

[46] See *Olafur Johannesson. Lög og réttur* (Law and Statutes). Reykjavik (1953), pp. 29—30.

[47] Reglementation de la navigation aérienne. Decret du Conseil des Ministres du mordad 1317 — 5 aout 1938 — available in *R. Aghababian* (Aghababoff).*Legislation Iranienne Actuelle interessant les etrangers et les Iraniens a l'etranger*. Teheran, 1939, pp. 145—148.

[48] *Eugène Pépin*, Development of the National Legislation on Aviation Since the Chicago Convention, in 24, The Journal of Air Law and Commerce (No. 1, Winter 1957), p. 9.

[49] Law of Civil Aviation 1949. 7 pages. Publication of the Department of Civil Aviation; Teheran Bank Melli Iran Press, (text in Persian and English translation) Reference taken from *Pépin, op. cit.,* p. 22.

[50] *Pépin, op. cit.*, p. 11.

[51] *Ibid.,* p. 10.

[52] Published in the official collection *Waqayi' al 'Iraqiya*, No. 1725 of August 16, 1939.

[53] In the official publication Government of Iraq, Ministry of Justice, Compilations of Laws and Regulations issued between 1st January 1939 and 31st December 1939. Baghdad, Government Press, 1941, pp. 91—94.

[54] Published in the *Waqayi' al 'Iraqiya*, No. 1729 of August 30, 1939. See English text in source quoted in footnote 2, p. 61.

[55] *Regio Decreto Legge 20 Agosto 1923, No 2207. Norme per la navigazione area.* (Royal Decree No. 2207. Provisions on Aeronautics of August 20, 1923); Official Gazette of the Kingdom of Italy No. 253, 1923; amended by Law of January 31, 1926, No. 1063; *ibidem* No. 111, 1926.

[56] *Regio Decreto 11 Gennaio 1925, No. 356. Approvazione del regolamente per la navigazione aerea* (Royal Decree of January 11, 1923, No. 356. Adoption of the Regulation for Aeronautics); *ibidem* No. 96, 1925; amended by Law of January 23, 1927, No. 325; *ibidem* No. 68, 1927; Law of May 4, 1928 n. 1946, *ibidem* No. 200, 1928; Law of May 13, 1928, No. 1555, *ibidem* No. 168, 1928; Law of October 31, 1929, No. 2486, *ibidem* No. 142, 1930; Law of April 11, 1932, No. 998, *ibidem* No. 197, 1932; Law of March 2, 1933, No. 673, *ibidem* No. 149, 1933; Law of December 18, 1933, No. 2348 (1934, 371); Law of December 3, 1934, No. 2106 (1935, 20); Law of March 25, 1935 n. 790 (Off. Gaz. of the Kingdom of Italy No. 133, 1935); Law of October 10, 1935, No. 2191 (1936, 6); Law of January 2, 1936, No. 360 (Off. Gaz. of the Kingdom of Italy No. 62, 1936); Law of November 25, 1937, No. 2361 (1938, 143); Law of April 15, 1938, No. 1350 (1669); Law of June 25, 1940, No. 1370 (Off. Gaz. of the Kingdom of Italy No. 239, 1940); Law of November 20, 1941, No. 1673 (1942, 547).

[57] *Regio Decreto 30 Marzo 1942 n. 327. "Approvazione del testo definitivo del codice della navigazione"* (Royal Decree of March 30, 1942, No. 327. Adoption of the Code of Navigation). *Codice delle leggi sulla navigazione maritima, interna e aerea* (Code of the Laws on Maritime Navigation, inland and aerial). Dott. A. Giuffre, Editor, Milan, 1952.

[58] Off. Gaz. of the Kingdom of Italy No. 107, 1923.

[59] Off. Gaz. of the Republic of Italy No. 131, 1948.

[60] *Pépin,* Development of the National Legislation on Aviation Since the Chicago Convention, in 24. The Journal of Air Law and Commerce (No. 1, Winter 1957), pp. 9—10, speaks of Jordan as a country which "at the beginning of 1957 still had pre-Chicago laws unaffected by the [Chicago] Convention".

[61] *Ibid.,* p. 18.

[62] *Budek Mieczyslaw,* Turkish Commercial Aviation, in 23 "The Journal of Air Law and Commerce" (No. 4, Autumn 1956), p. 439.

[63] *Ibid.,* p. 439, Note 82.

[64] La République Egyptienne. Ministère de la Guerre. Administration de l'Aviation Civile. *Recueil des Legislations Egyptiennes de l'Aviation, Lois, Décrets-lois, Décrets, Arrêtés Ministériels. Accords Bilatéraux. Conventions Internationales.* Le Caire, 1954, p. 893.

[65] Civil Aviation Law 1949. 21 pages. Publication of the Ministère des Travaux Publics, Service des Communications aériennes, Beyrouth, Imp. Ad. Dabbour, 1949. (Text in French, English, and Arabic).

[66] *Eugène Pépin,* Development of the National Legislation on aviation since the Chicago Convention, in 24 The Journal of Air Law and Commerce (No. 1, Winter 1957), p. 11.

[67] *Pépin,* Development of the National Legislation on Aviation since the Chicago Convention, in 24 The Journal of Air Law and Commerce, (No. 1, Winter 1957), p. 16.

[68] *Pépin,* Development of the National Legislation on Aviation Since the Chicago Convention, in 24 The Journal of Air Law and Commerce (No. 1, Winter 1957), p. 16.

[69] Liberian Code of Laws of 1956, Vol. III, Ithaca, New York, USA, Cornell University Press (1957), p. 1348—1349.

[70] *Eugène Pépin,* Development of the National Legislation on Aviation since the Chicago Convention, in 24 The Journal of Air Law and Commerce (No. 1, Winter, 1957), p. 16.

[71] *Frede Castberg, Norges Statsforfatning* (Norwegian Constitution), 2nd ed., vol. I, Oslo, 1947, 227, note 1, and *J. Andenaes, Statsforfatningen i Norge* (Government of Norway, Oslo, 1948), pp. 69—72.

[72] Text as of January 1, 1953 in *Norges Lover* 1682—1952 (Laws of Norway, 1682—1952, Oslo, 1953), pp. 1349—1358. An amendment of July 17, 1953 is published in *Norsk Lovtidende,* the Norwegian official gazette, 1953 No. 27, p. 789. English translation in Compilation of Norwegian Laws etc. for the Use of Foreign Service Representatives 1814—1953, Oslo, 1956, pp. 451—467.

[73] *Regler for fremmede krigsskibs og militaere luftfartøiers adgang til norsk territorium under fredsforhold,* confirmed by the King on August 19, 1938 (*Norsk Lovtidende* 1938, No. 32, p. 1168).

[74] Ratified by Norway on April 24, 1931. See *Overenskomster med fremmede stater* (Norwegian Treaty Collection) 1932, No. 1, p. 1.

[75] Norway ratified this Convention on May 2, 1947. *Ibid.,* 1948, No. 3, p. 376.

[76] Sec. 11 of the Rules for the Entry of Foreign Warships and Military Aircraft.

[77] *Lex cit.* Sec. 12.

[78] *Forskrifter for merking av civile luftfartøyer* of January 26, 1955, *Norsk Lovtidende* 1955, No. 30, p. 868.

[79] See *Castberg, op. cit.,* p. 227, and *Edvard Hambro, Norsk Fremmedrett* (Norwegian Law on Aliens, Oslo 1950), pp. 64—66.

[80] *Norges Lover,* p. 54.

[81] *Norsk Lovtidende* 1935, No. 33, p. 893. Amended by Resolution of December 10, 1937, *ibid.,* 1937, No. 46, p. 1303. English translation available in Compilation of Norwegian Laws, pp. 350—353.

[82] *Norsk Lovtidende* 1952, Part. 2, p. 824. English translation in Compilation of Norwegian Laws, pp. 354—357.

[83] *Eugène Pépin,* Development of the National Legislation of Aviation since the Chicago Convention, 24 J. of Air Law and Commerce (No. 1, Winter, 1957), p. 10.

[84] Ligue des États Arabes, Secretariat Général. Commission Permanente des Communications. La Sous-Commission pour las Affaires de l'Aviation Civile. *Rapport présenté par le President de la Sous-Commission, le Group Captain Ibrahim Hassan Gazzarine* (E.M.) in La Republique Egyptienne. Ministre de la Guerre. Administration de l'Aviation Civile. *Recueil des Legislations Egyptiennes de l'Aviation*. Le Caire, Imprimerie Nationale, 1954, pp. 878, 881.

[85] See *Robert Malmgren, Sveriges Författning; en lärobok i Svensk statsrätt* (Swedish Constitution; Textbook of Swedish Constitutional Law, Malmö, 1929), p. 24. See also *Kurt Grönfors, Grundzüge der schwedischen Luftfahrtgesetzgebung* (Principles of Swedish Aviation Legislation), Zeitschrift für Luftrecht (Berlin), Vol. I (1952), p. 45; and the *Förslag till brottsbalk, avgivet av Straffsrättskommittén* (Draft of the Criminal Code, submitted by the Committee on Penal Law), Stockholm, 1953, p. 446. As a general rule, the Swedish territorial waters extend to four miles from the coast. See *Torsten Gihl*, "The Limits of Swedish Territorial Waters", American Journal of International Law, Vol. 50 (1956), pp. 120—122.

[86] *Sveriges Rikes Lag* [General Code of Sweden], 1957 ed., pp. B 359—364.

[87] *Ibid.,* p. B 359.

[88] *Kungörelse med vissa bestämmelser rörande tillämpningen av förordningen den 26 maj 1922 (nr. 383) om luftfart av april 20, 1928.* Published originally in *Svensk Författningssamling,* the Swedish official gazette, 1928, under No. 85; a more recent text found also in *Ragnar Lindberg, comp. Försvaret* (Defense Legislation), Stockholm, 1954, pp. 606—609.

[89] *Kurt Grönfors, Om trafikskadeansvar utanför kontraktsförhållanden* (Tort Liability for Damages in Traffic Cases), Stockholm, 1952, p. 262.

[90] See also *Halvar G. F. Sundberg, Folkrätt* (International Law), Stockholm, 1944, pp. 104—106; and *Torsten Nylén,* Scandinavian Co-operation in the Field of Air Legislation, The Journal of Air Law and Commerce, Vol. 24 (1957), pp. 36—46.

[91] Published in *Svensk Författningssamling* 1957, No. 297 (June 20, 1957).

[92] All issues of *Svensk Författningssamling* up to and including the issue of December 6, 1957 (containing Law No. 664) were examined.

[93] *Eugène Pépin,* Development of the National Legislation on Aviation since the Chicago Convention, in 24 the Journal of Air Law and Commerce (No. 1, Winter, 1957), p. 9.

[94] Published in the *Journal Officiel* No. 66 of December 12, 1949. Reprinted by the Ministry of Public Works and Communications, Department of Civil Aviation (42 pages). Imp. of the Syrian Republic (Text in French and English). French text in 4 *Revue francaise de droit aérien,* (1950), pp. 37—50.

[95] The quotations from the law are a translation from the French text which appeared in the French Revue cited above in No. [94]. The official English text of this law, viz. the Law Regulating Air Transportation and Navigation 1949, published in the *Journal Officiel,* No. 66 of December 12, 1949, reprinted by the Ministry of Public Works and Communications, Department of Civil Aviation (42 pages), Printing Office of the Syrian Republic (text in French and English), was not available. *See Pépin, op. cit.,* p. 23.

[96] *Eugène Pépin,* Development of the National Legislation on Aviation Since the Chicago Convention, in 24 The Journal of Air Law and Commerce, (No. 1, Winter, 1957), p. 19.

[97] 23, The Journal of Air Law and Commerce (No. 4, Autumn 1956), pp. 379—478. An earlier treatise on Turkish Air Law in French is *Nasir Zeytinoglu's Étude de droit aérien turc".* Lausanne, 1951.

[98] The signatories to the Convention were Great Britain, USA, Sweden, France, Czechoslovakia, Netherlands, Denmark, Iraq, Greece, Lebanon, Norway, Jordan, Syria, Italy, Israel, Egypt, Brazil, Switzerland, Spain, and Iran. Budek, *op. cit.,* p. 438—439.

[99] *Budek, op. cit.,* p. 429, 430.

[100] Article 1 of the Convention on International Civil Aviation signed at Chicago December 7, 1944, ratified by the Turkish Grand National Assembly on June 5, 1945,

reads: "*Art. 1.* The Contracting States recognize that every State has complete and exclusive sovereignty over the airspace above its territory."

[101] Sec. 8 of Part II of this Act prohibits aircraft from flying over such areas in the United Kingdom as may be specified by an Order in Council.

[102] *Zakon o vazdusnoj plovidbi od 22 februara 1928* (Law of Aviation, February 22, 1928), *Sluzbene Novine* of the Kingdom of Serbs, Croats and Slovenes, No. 50-XIII, 1928; amended by the Law of January 14, 1930; *ibidem*, No. 18-VII, 1930.

[103] *Bertold Eisner, Medjunarodno privatno pravo II* (Conflict of Laws). Zagreb, 1956, p. 130.

[104] *Uredba o zracnoj plovidbi od 1 juna 1949* (Decree of Civil Aviation, June 1, 1949); Law No. 392; Official Gazette of the FPRY No. 47, 1949.

[105] *Zakon o obalnom moru Federativne Narodne Republike Jugoslavije od 1 decembra 1948* (Law on Territorial Waters, December 1, 1948); Law No. 876; *ibidem* No. 106, 1948.

[106] Convention for the Regulation of Aerial Navigation, Paris, October 1919. Ratified by the Law of December 6, 1926; *Sluzbene Novine; ibidem*, No. 8-II, 1927.

[107] *Ilija Przic, Osnovi Vazduhoplovnog prava* (Principles of the Law of Aviation); Beograd, 1926, p. 19.

[108] Decree of the Federal Executive Council of the FPRY of November 28, 1953 ratifying the International Civil Aviation Convention, Chicago, 1944. Official Gazette of the FPRY, Appendix — International Treaties No. 3, 1954.

[109] *Milan Bartos, Medjunarodno javno pravo* (International Law), Beograd, 1951, p. 106.

[110] *Bertold Eisner, op. cit.*, p. 151.

World Security and the Peaceful Uses of Outer Space

Eilene Galloway

The objective of the United Nations in establishing the Committee on the Peaceful Uses of Outer Space is to safeguard the right of peoples of all nations to beneficial results from space exploration. Attainment of this objective requires international cooperation which can be furthered by UN assistance for research, exchange and dissemination of information, encouragement of national research programs, and the study of legal problems arising from space exploration. Solutions proposed for legal problems can be evaluated in terms of whether or not they are likely to contribute to the betterment of mankind.

Organization for international scientific space activities exists in certain specialized agencies of the United Nations and in recognized non-governmental organizations. Space activities are highly diversified and whether the approach to their development is scientific or legal, the main problem is one of coordination of personnel, resources, facilities, rules, regulations, statutes, treaties, and agreements.

The state of the art of science and technology in space exploration will determine, in many cases, what controls are feasible and practicable. Scientists and engineers need to be informed of the impact of national and international laws upon the conduct of their projects. Lawyers formulating guidelines for the future need to keep abreast of fast-developing space sciences. The problem of culling the scientific facts essential to the solution of legal questions may be met by establishing a close working relationship between the IAF's International Institute of Space Law and the International Academy of Astronautics.

I. International Objective and Role of the United Nations

The Third Colloquium on the Law of Outer Space meets at a time when the International Institute of Space Law has been organized into working groups for the study of specific problem areas. The ideas we generate may lead to action for a system of world security wherein the people of all nations can be assured that space activities will be conducted for peaceful purposes. The challenge we face now is to study the legal problems, already so well identified, in conjunction with the developing facts of science and technology; in awareness of the need of policy-makers to know the advantages and disadvantages of alternative courses of action; and in recognition of the objective that space exploration should be undertaken for the benefit of mankind.

As we expand from general and theoretical studies of space law and plunge into the detailed research and analysis required for the solution of particular problems, it is essential to understand the existing international situation as it relates to outer space. Our present position has been shaped by an international objective, a pattern of organization and administration which may contribute to the desired result, and the dynamic force exerted by space science and technology upon international ideas and relations.

The international objective of space exploration is contained in the United Nations resolution "International Co-operation in the Peaceful Uses of Outer Space" which was adopted unanimously in a plenary session of the General Assembly on December 12, 1959. In establishing the Committee on the Peaceful Uses of Outer Space, the resolution states that the General Assembly,

> *Recognizing* the common interest of mankind as a whole in furthering the peaceful use of outer space,
> *Believing,* that the exploration and use of outer space should be only for the betterment of mankind and to the benefit of States irrespective of the stage of their economic or scientific development,
> *Desiring* to avoid the extension of present national rivalries into this new field,
> *Recognizing* the great importance of international co-operation in the exploration and exploitation of outer space for peaceful purposes,
> *Noting* the continuing programmes of scientific co-operation in the exploration of outer space being undertaken by the international scientific community,
> *Believing* also that the United Nations should promote international cooperation in the peaceful uses of outer space,
> *Establishes a Committee on the Peaceful Uses of Outer Space* [1].

In brief, the objective is that space activities should be conducted for the benefit of all people. The methods to be used in attaining this objective are implicit in the resolution: promotion by the United Nations of cooperation between nations and assistance for space programs of the international scientific community.

The directives given by the General Assembly to the Committee also indicate the role of the United Nations and the means of achieving international cooperation: UN assistance for research, exchange and dissemination of information, encouragement of national research programs, and the study of legal problems arising from space exploration. The Committee was also called upon to arrange an international scientific conference of interested members of the United Nations and the specialized agencies.

The objective and general methods of procedure are substantially the same as those adopted by the United Nations General Assembly on December 13, 1958 when the *Ad Hoc* Committee on the Peaceful Uses of Outer Space was established [2]. One difference in wording is that the 1958 resolution stated that it is the common aim that outer space "should be used for peaceful purposes *only*", whereas the 1959 resolution creating the permanent committee recognizes that the common interest of mankind would be served by "furthering the peaceful use of outer space ... (which) ... should be only for the betterment of mankind ..."

The present position of the United Nations has several points of significance for those concerned with space law.

First, the international declaration of policy seeks to safeguard the rights of the peoples of all nations to beneficial results from space exploration. This worldwide concept of human welfare must necessarily affect the character of controls for strengthening peaceful conditions and avoiding national rivalries which might lead to hostilities. A primary objective has therefore been provided whereby solutions proposed for legal problems can be evaluated in terms of whether or not they are likely to contribute to the betterment of mankind.

Second, problems concerned with outer space and with disarmament are being considered separately by the United Nations. This policy determination was made so that lack of agreement on an effective system of inspection and control of armaments need not delay progress in analyzing the unusual effects of space exploration upon international relations. A unique feature of the space age is that it began and is developing under auspicious patterns of cooperation established for the International Geophysical Year.

Third, the role of the United Nations will be further clarified after the Committee on the Peaceful Uses of Outer Space reports on its review of areas which "could appropriately be undertaken under United Nations auspices", including, *inter alia:*

> Assistance for the continuation on a permanent basis of the research on outer space carried on within the framework of the International Geophysical Year;
>
> Organization of the mutual exchange and dissemination of information on outer space research;
>
> Encouragement of national research programs for the study of outer space, and the rendering of all possible assistance and help toward their realization;
>
> And, furthermore, the Committee was requested;
>
> To study the nature of legal problems which may arise from the exploration of outer space [3].

Fourth, cooperation between nations and coordination of programs undertaken by organizations of the international scientific community are envisaged as elements of the pattern for space research and operations. This position is based upon the desire to continue, and to build upon, the organizational and administrative practices by which space activities have been, and are being, developed.

II. Organization of International Scientific Space Activities

The years of planning which preceded the International Geophysical Year laid a firm foundation for international cooperation on scientific projects which were worldwide in their scope. When the International Council of Scientific Unions (ICSU) established its Special Committee for the International Geophysical Year (CSAGI) in 1953, the forces set in motion carried through the period of the IGY (from July 1, 1957 to December 31, 1958) with such success that organization on a permanent basis was stimulated.

When the eighth General Assembly of the International Council of Scientific Unions met in Washington, D.C., from October 2—6, 1958, the Committee on Space Research (COSPAR) was provisionally established "to provide the world scientific community with the means whereby it may exploit the possibilities of satellites and space probes of all kinds for scientific purposes, and exchange the resulting data on a cooperative basis" [4].

The Charter for COSPAR's permanent organization was adopted at a meeting in Amsterdam on November 13, 1959. The provisions concerning membership and the election of officers are especially significant because of the key role performed by this international nongovernmental organization in planning and coordinating space research and development. National members and international scientific unions which adhere to ICSU are eligible for membership if they wish to participate in COSPAR and are actively engaged in space research. The Executive Council is composed of nine representatives of scientific unions, and a seven-member bureau operating under a President, two Vice Presidents who alternate in precedence, and four other members chosen from lists of names furnished by the vice presidents. The United States National Academy of Sciences and the Academy of Sciences of the USSR present separate nominating slates for the two vice presidential positions.

When COSPAR held its first meeting under the new Charter in Nice, France, from January 8—16, 1960, the election of officers resulted in the representation of nations identified with both the West and East. The President is a citizen of The Netherlands, the two vice presidents are from the United States and the USSR, while the four members elected to the Executive Council represent the United Kingdom, France, Czechoslovakia, and Poland.

COSPAR carries out its functions by means of working groups, now organized on the subjects of Tracking and Telemetering, Scientific Experiments, Data and Publications. In addition, COSPAR depends upon reports of national space activities, and at its January 1960 meeting received information on the status of twelve national programs concerned with space research and operations.

The relation of COSPAR to the United Nations is especially significant. Under an agreement made by ICSU, the United Nations Educational, Scientific and Cultural Organization (UNESCO) stimulates space research by promoting both international and regional arrangements. UNESCO supports COSPAR in its work, undertaking only those requests for space research which COSPAR cannot perform and those requiring intergovernmental agreement. By invitation, the President of COSPAR consulted with the United Nations *Ad Hoc* Committee on the Peaceful Uses of Outer Space, UNESCO's Department of Natural Sciences, and with the Administrative Radio Conference of the International Telecommunication Union. COSPAR sent an observer to the ITU conference which was held in Geneva from August 17 to December 31, 1959. This conference is noteworthy for having reached agreement on the terms of a treaty which includes provision for the allocation of radio frequencies to space vehicles. Both the United Nations and individual national governments depend upon COSPAR to perform functions of international coordination of scientific space activities.

In addition to UNESCO and the ITU, it can be anticipated that a number of other inter-governmental organizations will be concerned with various aspects of outer space research and development, *e.g.*, the World Meteorological Organization (WMO), the International Civil Aviation Organization (ICAO), the International Atomic Energy Agency (IAEA), the World Health Organization (WHO), and the Inter-Governmental Maritime Consultative Organization (IMCO) [5].

Harmony between the objectives and organization of national and international programs is vital for successful coordination. The policy objectives and organization of the United States for international space activities are in consonance with those of the United Nations and the world scientific community.

By unanimous vote the Congress of the United States passed a resolution declaring:

That it is the sense of Congress:

> *That the United States should strive, through the United Nations or such other means as may be most appropriate, for an international agreement banning the use of outer space for military purposes;*
> *That the United States should seek through the United Nations or such other means as may be most appropriate an international agreement providing for joint exploration of outer space and establishing a method by which disputes which arise in the future in relation to outer space will be solved by legal, peaceful methods, rather than by resort to violence;*
> *That the United States should press for an international agreement providing for joint cooperation in the advancement of scientific developments which can be expected to flow from the exploration of outer space such as the improvement of communications, the betterment of weather forecasting, and other benefits; and*
> *That the Congress respectfully requests the President to effectuate in every way possible the objectives set forth in this resolution [6].*

World Security and the Peaceful Uses of Outer Space

In establishing the National Aeronautics and Space Administration, Congress declared that "it is the policy of the United States that activities in space should be devoted to peaceful purposes for the benefit of all mankind". The law further provides that the aeronautical and space activities of the United States shall be conducted so as to contribute materially to "cooperation by the United States with other nations and groups of nations in work done pursuant to this Act and in the peaceful application of the results thereof". Implementation of this policy is found in the provision that

> *The Administration, under the foreign policy guidance of the President, may engage in a program of international cooperation in work done pursuant to this Act, and in the peaceful application of the results thereof, pursuant to agreements made by the President with the advice and consent of the Senate* [7].

Informal arrangements for cooperative programs may also be undertaken by NASA.

In accordance with these policy directives, NASA cooperates with the United Nations and, working through the US National Academy of Sciences, with the Committee on Space Research (COSPAR). An Office of International Programs has been established by NASA to develop cooperative activities.

International NASA programs that are now under way fall into four main categories: the worldwide network of tracking and telemetry stations so essential for the acquisition of data from satellites and space probes; joint programs with scientists and engineers of nations cooperating in space research; exchange of scientific and technical information, making data available for evaluation to scientists throughout the world; and training programs and exchanges with foreign scientists.

These programs are being conducted in accordance with a policy of freedom and openness which is truly international in matching the scientific and technological requirements of the space age. A case in point is the offer by the United States to the Soviet Union of the use of the special tracking facilities designed for the man-in-space program, Project Mercury. Dr. T. Keith Glennan, Administrator of NASA announced that

> *As an evidence of our interest in international cooperation, we would be most happy to offer the services of our tracking network in support of the scientists of the Soviet Union when and if that nation undertakes a manned space-flight program. Data could be acquired and transmitted in its raw state to the Academy of Sciences in Moscow. A precedent for this sort of thing has been established in the IGY operation when the United States supplied to the Soviet scientists, as of July 1959 some 46 tape recordings of Sputnik I, II, and III. Should special recording or data read-out equipment be required, I am sure that we would be happy to provide them or to utilize equipment furnished by the Soviet scientists. In such a cooperative venture we could help them to keep in continuous or essentially continuous contact with their astronaut* [8].

Another example of the unity of effort which is sought by NASA in generating international cooperation in space research is the offer to launch for other countries individual experiments or complete scientific payloads in artificial earth satellites.

NASA authorized the delegate from the National Academy of Sciences to make this proposal at the second meeting of COSPAR held at The Hague in March 1959. Scientists from a number of countries are interested in the opportunity of developing their space projects in cooperation with the United States. Agreements on joint programs of this type have already been reached with the United Kingdom, Canada, and Italy, and are in the process of being worked out with scientists in additional countries.

III. Scientific and Technological Facts of Space Exploration

The most striking fact about the advent of the space age is the rapidity of the pace of scientific and technological development since the first earth satellite was orbited on October 4, 1957. The rate of growth in space exploration may not seem great to those who compare present achievements with distant scientific goals as yet unattained. But by practically any other criterion of measurement the accomplishments made by scientists and engineers, in substantially less than three years, are impressive.

The urgency with which activities are being expanded into the space environment is revealed by the numbers of successfully launched satellites and space probes. Between October 1957 and August 12, 1960, 28 space vehicles were sent into orbit around the earth; 3 were launched into solar orbits; and one lunar impact was made. On August 12, 1960, there were 13 earth satellites still in orbit; 3 space vehicles in solar orbit; and 7 of these devices were still transmitting information from outer space to the earth. The international summary of satellites and space probes was published by the National Aeronautics and Space Administration [9].

Current Summary (August 12, 1960)

Earth Orbit: USA 12
               USSR 1

Solar Orbit: USA 2
              USSR 1

Transmitting: USA 7
              USSR 0

Complete Summary (Successfully launched to date)

Earth Orbit: USA 23
               USSR 5

Solar Orbit: USA 2
              USSR 1

Lunar Impact USSR 1

Information on the distances from earth reached by orbiting space vehicles affords a factual basis for considering controls that might be applicable to spatial areas and spacecraft. Eighteen satellites established a perigee lower than 300 miles, one of the figures theoretically suggested as a demarcation between airspace and outer space. Of these 18 satellites, 10 had a perigee of 156 miles or lower. Going into an elliptical orbit, many satellites come comparatively close to the earth for a few minutes, only to be thousands of miles away a short time later. For example, the US Explorer I satellite, with an estimated lifetime of 3 to 5 years from its launching date (January 3, 1958) has a perigee of 224 miles and an apogee of 1 573 miles. Explorer VI, with a life expectancy of 1 year from August 7, 1959, has a perigee of 156 miles and an apogee of 26 357 miles.

This type of scientific information raises questions concerning definitions which have been offered in the hope of clarifying national sovereignty in airspace and outer space. Is the space vehicle, or space itself, to be controlled? Some idea of the complexities of this question can be gleaned from considering an example such as Lunik III, a Soviet translunar earth satellite, which had an elliptical orbit of 625 000 miles for 15 days. Another example is that of the US Pioneer V space probe, launched on March 11, 1960 with an expected lifetime of 100 000 years, a perihelion of 74 967 000 miles and an aphelion of 92 358 000 miles. Jodrell Bank in England received information from Pioneer V from a record distance of 22.5 million miles from the earth [10].

Time as well as distance must be considered in establishing the necessary correlation between scientific facts and proposals for the solution of legal problems. While the time required to complete an orbit naturally varies with different satellites, many have been circling the earth in approximately 1½ to 2 hours. The factor of speed may be controlling in determining the feasibility of some of the proposed controls.

The nature and amount of information being obtained from scientific exploration constitutes another criterion by which achievements may be measured and evaluated. Several examples will illustrate the yield which can be anticipated from a study of research sources which are easily available to the international legal profession.

Information from the US Explorer I satellite, launched on January 31, 1958, led to the discovery of the radiation belt around the earth. From the solar-powered signals of Vanguard I, launched by the United States on March 17, 1958, with a life-expectancy of 200 to 1000 years, the shape of the earth and the locations of Pacific Islands are being measured more exactly. Great strides have been made in communications satellites. The Project Score satellite (December 18, 1958) transmitted from outer space a recorded message from President Eisenhower. The US Pioneer V space probe achieved several scientific "firsts" in transmitting 138.9 hours of data from the space between the orbits of Venus and the earth. Outstanding results from weather experiments have been recorded from the US satellites Explorer VII (October 13, 1959) and the Tiros television and infra-red observation satellite (April 1, 1960). Circling the earth at a speed of 99.19 minutes, Tiros relayed 22 952 cloud cover photographs which are considered highly successful.

The Soviet Lunik I (January 2, 1959) went into orbit around the sun on a 15-month cycle. Lunik II (September 12, 1959) impacted the surface of the moon with scientific instruments and the Soviet coat of arms. Lunik III (October 4, 1959) took high precision photographs of 70 percent of the dark side of the moon and transmitted them back to earth [11].

Obviously, this paper can do little more than indicate the main categories of facts which are available for study, and emphasize the necessity for a thorough evaluation of the impact of scientific knowledge upon problems requiring legal measures for their solution.

IV. Implications for the Analysis of Legal Problems

Highlights of the existing international objective, organization, and scientific facts have been provided as a background for the analysis of legal problems arising in space exploration. As the working groups of the International Institute of Space Law begin their studies of the past and present in order to formulate guidelines for the future, it is necessary, also, to consider the assumptions which may underlie their work.

For purposes of discussion, we may consider the validity of the following basic assumptions. Most of the legal problems concerning outer space have been identified, but intensive research and analysis are now required for the solution of specific questions. The international legal profession must find ways of keeping abreast of the rapid development of space science and technology so that legal controls may be logically correlated with space activities. Space law, now in its earliest stage of development, is a stimulating challenge to non-governmental working groups and individual thinkers who have an opportunity for creative thinking in analyzing proposed objectives, assumptions, and solutions of legal space problems. We have entered a period when policy decisions are in the process of being made by officials of national and international political and scientific organizations. Private groups and individuals are free to think of the alternatives by which specific questions may be mitigated or solved, and to offer the policymakers, at a critical time, some well-reasoned estimates of the probable consequences of each course of action.

Within the context of such terms of reference, what are the implications for space law of the United Nations resolution embodying an international objective for space exploration? The objective, that mankind has a common interest in outer space and its peaceful exploration for his betterment, constitutes a standard by which the probable effects of national and international actions can be measured. Every proposal could be analyzed as to whether or not it might contribute to the betterment of mankind. Proposals affecting national sovereignty or international control could be tested for their likelihood of advancing or retarding benefits for the people of all nations. Particular theories of national sovereignty, concerned with dividing jurisdictions according to zones of altitude or cones of space or free areas, would face the testing question of whether their adoption would help or hinder international cooperation in the peaceful uses of outer space. In terms of achieving the objective, what are the advantages and disadvantages of a system of controls for defined areas of space as contrasted with controls primarily designed to apply to space vehicles? Another question to which little attention has been paid is the legal implication of the growing scientific concept of aerospace in which there is no differentiation between airspace and outer space.

A problem of significance to those concerned with formulating laws for outer space is whether that area is to be used only for peaceful satellites or for both peaceful and military space vehicles. It would be necessary, of course, to define what is meant by "peaceful" and "military". But beyond that, two alternative approaches to law may be envisaged: one would be concerned with permitting peaceful and prohibiting military spacecraft; the other would consist of rules for the regulation of both civilian and military space vehicles. A better understanding of the implications of this situation could result from legal studies undertaken by the International Institute of Space Law.

We may next inquire concerning the implications for space law of the organization for space exploration established by the United Nations and by the international scientific community.

International cooperation rather than international control is now called for in the resolution establishing the United Nations Committee on the Peaceful Uses of Outer Space. There is no official suggestion for a supranational agency to control or conduct the space programs of nation states. Nor is there any official development toward an outer space organization comparable to the International Atomic Energy Agency. The possibility of such developments might be studied by individual lawyers intent upon exploring the pros and cons of different types of organizational arrangement. In the absence of a centralized world authority over space activities, we may expect cooperation between nations to be achieved by the normal method of bilateral and multilateral treaties and agreements.

The specialized agencies of the United Nations and the organizations established by the international scientific community will also be used for the necessary coordination of space activities. From this practice, functional controls may be expected to develop over such areas as weather prediction, the allocation of radio frequencies, and navigation. Many specialized agencies already have rules and regulations which may apply in some degree to space activities.

The problem created for international law is how to coordinate the administrative regulations of various agencies concerned with specific functions of space exploration. To what extent need such rules be made uniform? Will international administrative regulations be coordinated by COSPAR, by the International Council of Scientific Unions, or by the United Nations Committee on the Peaceful Uses of Outer Space? Space activities are highly diversified and whether the approach is scientific or legal, the main problem is one of coordination — coordination of personnel, resources, facilities, rules, regulations, statutes, treaties and agreements.

The state of the art of science and technology in space exploration will determine, in many cases, what controls are feasible and practicable. What can and cannot be done technologically may determine what can and cannot be done legally. But it is also possible that the application of certain historic legal concepts to space activities might preclude courses of action which are technologically possible.

We need to establish a bridge between the legal and scientific professions in order to ensure their coordination. Scientists and engineers need to be informed of the impact of national and international laws upon the conduct of their projects. Lawyers who are formulating guidelines for the future need to keep abreast of the developing facts of space science and technology. There is a tremendous outpouring of scientific data, not all of which is pertinent to problems of law and jurisprudence. The problem of culling out those facts which are essential to the solution of legal questions may be met by establishing a close working relationship between the International Academy of Astronautics and the International Institute of Space Law. By mutual effort and cooperative procedures we shall be able to make a contribution to world security through the peaceful uses of outer space.

References

[1] United Nations Document A/RES/1472 (XIV) (A/C. 1/L/247, as amended). The membership of the Committee consists of Albania, Argentina, Australia, Austria, Belgium, Brazil, Bulgaria, Canada, Czechoslovakia, France, Hungary, India, Iran, Italy, Japan, Lebanon, Mexico, Poland, Romania, Sweden, the Union of Soviet Socialist Republics, the United Arab Republic, the United Kingdom of Great Britain and Northern Ireland, and the United States of America.

[2] United Nations Document 1348 (XIII), December 13, 1958. *See also* United Nations Report of the *Ad Hoc* Committee on the Peaceful Uses of Outer Space. A/4141 76 p. (July 14, 1959).

[3] United Nations Document A/RES/1472, XIV (December 17, 1959).

[4] United Nations Document A/4141, p. 10 (July 14, 1959).

[5] *Ibid.,* p. 15—26.

[6] House Concurrent Resolution 332. United States House of Representatives, June 2, 1958; United States Senate, July 23, 1958.

[7] The National Aeronautics and Space Act of 1958, United States Public Law 85-568, 85th Congress. July 29, 1958.

[8] *T. Keith Glennan,* Opportunities for International Cooperation in Space Exploration. Address made before the World Affairs Council, Pasadena, Calif. (December 7, 1959).

[9] International Satellite and Space Probe Summary. National Aeronautics and Space Administration. Office of Public Information NASA Rel. No. 60-244 (August 15, 1960).

[10] NASA Authorization for Fiscal Year 1961. Hearings before the NASA Authorization Subcommittee of the Committee on Aeronautical and Space Sciences, United States Senate. 86th Congress, 2nd session on H.R. 10809. June 30, 1960. Part. 2. Scientific and Technical Aspects of NASA Program. United States and Russian satellites, lunar probes and space probes, 1957 to June 1960, pp. 987—1006.

[11] *Ibid.*

International Space Law and Outer Space

Thos. E. Martin

There are difficulties in establishing a recognized boundary between air space and outer space. Should the division be defined in terms of the physical characteristics of the air? Or should it be based upon the purpose of flight or on the physical characteristics of flight-craft? There are, for example, vehicles which have the characteristics of both aircraft and spacecraft, operating on aerodynamic principles in one part of their flight and on pure rocket principles in another. All problems would not be solved by fixing a stationary boundary. If a relatively high altitude were fixed upon, artificial satellites could orbit inside and outside the area, depending upon the apogee and perigee. It would make little sense to impose one legal regime or status on the satellite at perigee and another on the same satellite at apogee. Another problem is that as altitude increases, the relevance and adequacy of normal air space boundaries decrease, and there is less relationship between objects located or activities taking place "above" specific national territory on the earth's surface. The search for appropriate criteria should include consideration of other factors than altitude, *e.g.*, the trajectory of space vehicles, flight mission, instrumentation, other functional characteristics.

No nation has objected to the launching of space vehicles, and thus it would seem there has been acceptance of the principle of freedom of exploration and scientific observation in much the same manner as was agreed upon in the Antarctic Treaty. Why should not serious consideration be given to adherence to such a practice with respect to celestial bodies in outer space?

With the growth of customary practice and the accumulation of laws and agreements dealing with particular subjects, a system of laws governing human relations in outer space will gradually be filled out.

I am pleased and honored to have been invited to speak before you today. The International Astronautical Federation has, since its foundation in 1950, provided a valuable international forum for the review of outer space problems and for the stimulation of action in governmental and international public bodies. I hope it will continue to provide this valuable service. In view of the uncharted nature of much of the work in astronautics, the contribution of private thinking and analysis can be of immense importance to the future development of man's efforts in outer space.

The stirring era in which we live is witness to the extension of man's activity into a medium, entirely new to him. In the realm of space exploration, man is faced with technical and scientific demands of unparalleled difficulty. At the same time, he is given a priceless opportunity — a chance to establish new relations with his fellow man in a new environment, relatively free from the limiting influence of precedent.

Starting with a clean slate, man should so plan his activities in outer space as to preclude the possibility of the armed conflicts and controversies which have characterized his history on this planet. He can promote a new atmosphere, based on scientific cooperation and the rule of law, which might serve as an example for his earth-bound relations with his fellow man. As Ambassador Henry Cabot Lodge stated in an address to the 14th General Assembly:

"International cooperation in the exploration of outer space offers an avenue along which nations may approach mutual understanding and peace. Working together on the great challenges of explorations beyond the confines of earth can create a new perspective, in which national boundaries and national rivalries recede in importance."

Certainly we can agree that the incentives for international cooperation in outer space are many and very substantial. Tremendous expense is involved in space exploration. Geographically widespread and yet tightly coordinated observation stations are required. Coordination of operations is imperative, if we are to avoid mutual interference and minimize the danger of accident and damage as traffic in outer space increases in intensity. All these factors urge the importance of international cooperation.

And there is another grim argument for cooperation: Scientific progress has inevitably made possible the use of outer space for new and more dangerous means of waging war. This fact requires that maximum cooperative efforts be made among the nations to guard against the use of outer space for aggressive purposes. Recognizing the great dangers lurking in such use, the United States over three years ago proposed a study of means to reserve outer space for peaceful purposes only. Today the United States stands prepared to enter upon such a study separately from the overall problem of disarmament. A concrete proposal in this direction has been made by the five Western powers to the Geneva Conference of the 10-Nation Committee on Disarmament. These nations have urged the banning of weapons of mass destruction in orbit or stationed in outer space.

I hope very much that outer space can be reserved as a great area for peace. It would be a tragic thing indeed if the wars and national rivalries, which have too often characterized relations among the nations here on earth, were to be projected into outer space.

It is gratifying to note that the strong incentives for international cooperation in outer space which I have mentioned have not been without effect. Even in the initial stages of space exploration, such cooperation has made an important contribution in the tracking of earth satellites and the assembly of scientific data. Wide recognition of the desirability of mutual assistance in this new field is reflected in the intelligent and significant cooperation of states, organizations, and individuals during the International Geophysical Year (1957—58). It is also reflected in the continuation of these activities under the aegis of the Committee for Space Research of the International Council of Scientific Unions.

A special, and important example of international cooperation, is furnished by the transmission by the United States to the USSR of a number of tape recordings of the data received from Sputniks I, II and III.

As the pace of outer space activities increases, there will grow, in addition to the need for cooperation, a need for regulation and control of such activities in order to minimize conflicts of interests and operations. This involves the establishment of broad principles and specific regulatory measures to meet operational requirements.

Because of the very newness of the field of outer space exploration, the context and nature of many potential problems are not yet clear. I therefore believe that it is premature, at this time, to attempt to draw up a comprehensive legal code to cover all contingencies in outer space. It is, after all, a cardinal rule that the development of international law must follow the dictates of concrete need. We do not know enough about the difficulties that may be encountered in outer space to foresee the precise nature of the entire range of regulatory measures which may one day be necessary.

There are, however, certain problems which can be identified at present and studied with regard to possible regulatory and control measures. In this task the United Nations *Ad Hoc* Committee on the Peaceful Uses of Outer Space has played an appropriately leading role. The report of this committee, published last year, made preliminary identification of a number of areas which call for international coordination and control on a priority basis.

Let me recall briefly what these areas are. Allocation and control of radio frequencies were listed by the *Ad Hoc* Committee as fundamentally important to space activities because of the exclusive dependence on radio communications for the transmission of scientific data and other messages to and from space vehicles. Coordination and controlled effort are necessary in order to avoid harmful communications interference between various space operations. Another matter — the identification and registration of space vehicles — is closely linked with the effort to maintain order in man's activities in outer space, as is also the coordination of launchings of space vehicles.

Still a fourth field of obvious importance concerns the avoidance of interference among space vehicles and between space vehicles and aircraft. Yet another concerns the reentry and landing of space vehicles. This would encompass the establishment of standards for the marking and identification of space vehicles as well as standards for the return of equipment and personnel from the territory of foreign states.

A final matter requiring priority attention was listed by the *Ad Hoc* Committee. This concerns the liability for injury and damage caused by space vehicles. Procedures will have to be set up for the determination of the fact and extent of liability in case of damage caused by space vehicles. In this connection, the United Nations Committee suggested that early consideration should be given to securing agreement that claims disputes between states will be submitted to the compulsory jurisdiction of the International Court of Justice. Safety standards should also be agreed upon with regard to notification of launching of vehicles, policing of areas of danger on the high seas and installation of safety equipment on missiles to ensure harmless destruction in the event of misfiring.

Now 1 would not go so far as to suggest that we install stop and go lights in outer space in order to control traffic there. But just as some regulatory measures become necessary for traffic on our international highways and on the high seas, so a certain degree of regulation and control will be essential with respect to outer space activities.

Subsequent to the recommendations of the *Ad Hoc* Committee, the International Administrative Radio Conference of the International Telecommunication Union took action at the initiative of the United States to allocate radio frequencies for space services. This constituted the first concrete measure of a regulatory character taken in the outer space field. As such, it represents an essential step toward establishment of an international basis for the orderly conduct of outer space activities.

The need for control in these various areas is obvious. It is interesting to note, in studying the matters involved, that there is a useful connection between identification and registration of space vehicles, advance notice of their launchings and a possible system for guarding against surprise attack or use of outer space for aggressive purposes. It may well be that experience gained in this field may be useful in disarmament and arms control measures.

In addition to these rather specific matters, the *Ad Hoc* Committee dealt with certain broader concepts. Among these, I might mention the applicability of international law to outer space, the problem of determining where outer space begins, the question of freedom of exploration and use of outer space, and questions relating to the exploration of celestrial bodies and assertion of territorial claims.

In the first instance, the committee pointed out — and I agree — that the provisions of the United Nations Charter are not limited in their operation to the confines of the earth. The committee expressed the hope that in accordance with Article II, Paragraph 1, of the United Nations Charter, activities in outer space will be conducted in recognition of the common interest of mankind in outer space. It emphasized that there should be respect for the common aim that outer space should be used for peaceful purposes only.

The committee also mentioned the unlimited applicability of the statute of the International Court of Justice. I would say that this is of tremendous importance, for it is absolutely essential that disputes between nations over rights and agreements concerning outer space be settled in a peaceable manner, without resort to force or violence.

In addition to general provisions for international cooperation and peaceful resolution of disputes, there are many specific legal procedures and principles governing air space and the high seas which might be adaptable to the treatment of problems arising from outer space activities. These precedents may afford some useful suggestions regarding space flight, exploitation of minerals and energy resources, and the maintenance of communications. Decisions on the registration of space vehicles may also be facilitated by examination of similar efforts that have been made in air and maritime law.

Regarding freedom of outer space for exploration and use, the *Ad Hoc* Committee took a liberal and forward looking view. It pointed out that "during the IGY .... and subsequently, countries throughout the world proceeded on the premise of the permissibility of the launching and flight of space vehicles .... regardless of what territory they pass 'over' during the course of their flight ....".

This premise appears to have been supported by the fact that such space activity has been undertaken and that no nation has raised objection to the launching of space vehicles by another. Thus it would seem to me there has been acceptance of the principle of freedom of exploration and scientific observation in much the same manner as was agreed in Washington last December 1 with respect to the continent of Antarctica.

A similar possibility exists regarding the exploration of and assertion of claims to celestial bodies. Last year agreement was reached by the signatory states to the effect that no territorial claims are to be submitted under the Antarctica Treaty. Why should not serious consideration be given to adherence to such a practice with respect to bodies in outer space? Certainly at the present state of man's knowledge and capabilities it would seem that the resources of natural bodies in space, like the vast regions of space itself, represent sharable assets of the community of nations. Scientific exploration of a planet conducted under the national auspices of one country should in no way involve the prohibition of similar, non-interfering, exploration by other countries.

The question of the establishment of a recognized boundary between air space and outer space is an especially complex one. In my estimation, it presents a surprising number of difficulties — difficulties which may possibly prove insuperable. The main argument for determining a space boundary is that it would help to preclude states from making claims to "sovereignty" over large parts of space on the assertion that they are "air space" rather than "outer space".

The criteria for such a boundary determination, however, are not easy to establish. Should the division between air space and outer space be defined in terms of the physical characteristics of the air? Or should it be based upon the purpose of flight or on the physical characteristics of flightcraft? There are, for example, vehicles which have the characteristics of both aircraft and spacecraft. These operate on aerodynamic principles in one part of their flight and on pure rocket principles in another.

In order to avoid these difficulties, it has been suggested that an arbitrary limit might be chosen. But where would the line be drawn? Would it be at an altitude of 50 000 feet, or 70 000 feet, or perhaps as high as a manned plane might fly? The result of any arbitrary limit, it seems to me, would likely be either to fetter space activities by inappropriate rules or to interfere unnecessarily with the existing regime of international aviation.

Even if the difficulties of fixing a stationary boundary were overcome, the achievement would not solve all problems. This would be most obviously the case were the boundary to be fixed at a relatively high altitude. For example, artificial satellites often come much closer to the earth at some points in their orbits than at others. In some cases the perigee, or point nearest the earth, falls within one or more boundaries which have been proposed, while the apogee, or farthest point, falls outside. To me, it would make little sense to impose one legal regime or status on the satellite at perigee and another on the same satellite at apogee.

Another problem would arise in connection with establishment of a high altitude space boundary. As altitude increases, the relevance and adequacy of normal air space boundaries decrease. There is progressively less relationship between objects located or activities taking place "above" specific national territory on the surface of the earth. For example, reentry and landing of an object traveling at high speed and high altitude are apt to take place hundreds, or even thousands, of miles farther on in its line of travel.

Rather than seek to limit space activity on the basis of altitude alone, other factors might be equally relevant. Should not one consider the trajectory of the space vehicle or object, its flight mission, the instrumentation and other functional characteristics of the vehicle in question, for example, in the search for appropriate criteria? Even with the resort to such criteria, however, much will remain in dispute. It would, I think, be ill-advised to accept a boundary for outer space before its practicality and utility are thoroughly established.

These are some of the imponderables with which we are faced in the formulation of legislation and procedures governing man's activities in outer space. The factors which must be taken into consideration are exceedingly complex, and the technical and scientific information as yet available concerning many aspects of such activities is far from complete. These facts underline the need for considerable caution in drawing up legal provisions relating to outer space activities.

Despite this complexity, however, the accelerating rate of outer space exploration renders indispensable the early consideration of regulatory measures in certain specific areas, as I have indicated earlier. With the growth of customary practice and the accumulation of laws and agreements dealing with particular subjects, a system of laws governing human relations in outer space will gradually be filled out.

In the growth of such a system, the international community will wish to see its basic policy aims, as affirmed in such documents as the United Nations Charter, reflected in the body of outer space law. These aims include the reservation of outer space for peaceful purposes, the encouragement of international cooperation and the encouragement of scientific research, particularly in the interest of achieving practical benefits to increase human welfare.

The hope of cooperation in outer space was put eloquently not long ago by Dr. T. Keith Glennan, Administrator of the National Aeronautics and Space Administration, when he said:

> *"Out of the efforts of the dedicated and inspired men of all nations may yet come that common understanding and mutual trust that will break the lockstep of suspicion and distrust that divides the world into separate camps today."*

Faced with the tremendous opportunities presented by outer space, we owe it to ourselves and to all posterity to rise to the occasion and meet the challenge with statesmanship and breadth of vision.

Two Problems of Outer Space Control: The Delimitation of Outer Space, and The Legal Ground for Outer-Space Flights

Vladimír Kopal

I. Complete and exclusive territorial sovereignty of states comprising inviolability of their relevant airspace, although it reaches much higher than aviation is developed today, applies only to a part of the space above the state territory. Neither the density of air, the height of the atmosphere surrounding our Earth and the reach of terrestrial gravity, nor the possibility of effective control can be decisive for the determination of its frontiers. The principal reason here is the fear from the abuse of the space contiguous to the Earth against the existence, independance and inviolability of the subjacent states, *i.e.* the viewpoint of security. This criterion is, however, a political concept, created by the state of international relations, which is different under disarmament and during an arms race.

II. There does not exist any legal norm forbidding the flights into outer space and in it for peaceful purposes. It is possible to require within the framework of future regime of space the validity for the principle that outer space is and should remain freely accessible under equal conditions to peaceful research and exploitation by all countries.

III. The originating legal regime of outer space must grow from the principles of peaceful coexistence of nations, as they regulate the relations between states and their nationals, be they realized on the land and sea, in the air, or in the outer space by space vehicles. The principle of sovereignty radiates into the outer space: freedom of access into the outer space and flights in it does not mean an unrestricted freedom of action, which would be *e.g.* directed against the integrity and territorial inviolability of subjacent states. Therefore the standpoint of security must be also applied to the determination of the purposes which astronautics can follow, especially in the part of outer space contiguous to the zone of complete and exclusive sovereignty.

Demilitarization and neutralization of outer space would remove obstacles to the solution of many legal problems. It would open the way also to the delimitation of outer space, favourable to the needs of the future astronautical development. In this respect the importance of the efforts to achieve the general and complete disarmamament which in a number of complex provisions should comprise complete prohibition and liquidation of all carriers of nuclear arms as well as these arms themselves, under an appropriate and effective control, is still more outstanding.

I.

One of the most important questions, which belongs to the large group of problems concerning the legal control of outer space, is the well known problem of how to delimitate the outer space and to find the real legal ground, on which the peaceful flights into outer space are based.

I consider as correct the opinion that *complete and exclusive territorial sover-*

eignty of states, comprising inviolability of their relevant airspace, incorporated in international air conventions and air legislature of states, although it reaches much higher than aviation is developed today, applies only to a part of the space above the state territory [1].

By the penetration of rockets into the cosmic space a new problem has been opened, which the air law, by which only the legal regime of the airspace was regulated, does not solve.

The affirmative answer to the question, whether the frontiers of sovereignty do exist or not, carries with it, of course, another question, *where* these frontiers lie, or to put it more exactly, *where* to place these frontiers, as we are setting out from the conviction that the legal regime of cosmic space is only being created and will be created. The answer to this question, so far published in literature, has been supported by various criteria. The low density of air, preventing the flight of aircraft, is the lowest of them. The height of the atmosphere, surrounding our Earth, is another one. The reach of terrestrial gravity is a still more distant criterion. Contrary to these criteria, setting out from natural and physical values, another criterion — a possibility of effective control — is based on technical progress, reached by the people, thus representing a considerably mobile frontier. It seems this mobility will reach in the near future such upward trend that, if the answer to the given question was setting out only from this criterion, it could mean after a time the same as the theory of sovereignty without any limits.

None of the above-mentioned criteria can be decisive for the question of accurate determination of the frontiers of territorial sovereignty.

The reason for which the states in our time affirm their sovereignty over the airspace is indeed not the fact itself that it is characterized by certain density of air, or that it is penetrated by territorial gravity, or finally, that they themselves are able to control this space up to a certain limit. The decisive reason here is *the fear from the abuse of this space by other states against the existence, independance, inviolability, and prosperity of the subjacent state, i.e. the viewpoint of security* [2].

The criterion of state security is, however, a political concept, created by the general state of international relations, which is different in conditions of general and complete disarmament on the one side and during an arms race on the other side. It stands to reason that the establishment of a firm legal norm on the upper limit of States sovereignty cannot be dependent on fast changing conditions. At the same time it cannot be and even will not be a solution once for all.

In respect to the expected difficulty of setting the most suitable frontier it has been proposed to *establish between these lines an intermediate zone* which would either be outside the reach of the sovereignty of the subjacent state, but should offer it certain advantages, or, on the contrary, would be subject to this sovereignty in principle, but a peaceful exercise of certain rights would be permitted in it to other states [3].

In principle such a solution meets to a considerable extent both pressing, but in this case contradictory, aspects: the security of states on the one hand, which in the given situation requires that the range of sovereignty be as great as possible, and the needs of astronautics which would require on the contrary that a frontier of complete and exclusive sovereignty set too high should not be an insuperable obstacle to it. Simultaneously it would enable that only the second of these lines, the frontier of the contiguous zone, where sovereignty to a certain extent reaches and of the outer space, which in principle would remain free, be subject more often to revision, considering the further turbulent, scientific and technical development and the requirements of security of the States. The tri-zone solution, in which the second line would be at a greater distance, also would bring certain advantages to the states the territory of which is not extensive enough that a rocket launched by them, could leave territorial space or return to it without infringing the sovereignty frontiers of the second state.

Of course, against the institution of the contiguous zone objections have already been expressed by some authors, referring to the difference of natural conditions [4], the aims, followed by a similar sea zone [5], *etc.* These differences, however, are not serious and do not exclude the existence of the contiguous zone with a different purpose as well as contents in the outer space.

On the other hand under the existing conditions of international relations, when serious infringements of States sovereignty in their airspace are taking place, when no result has yet been achieved in disarmament efforts, when coordination and cooperation in launching space vehicles are missing, it would be quite difficult to exploit practically the advantages of the intermediate zone which would require an effective control over the maintenance of the purposes which it should serve. Under these circumstances the States will evidently prefer that their full and exclusive territorial sovereignty should reach to the most distant line.

II.

The first satellites of the Earth, as it is known, were launched within the framework of the International Geophysical Year which after its expiration on December 31, 1958, was extended to December 31, 1959, as International Geophysical Cooperation. No state has protested with reference to its territorial sovereignty either against the announced programmes, or after the launching of the first satellites, nor against further launching of rockets after the conclusion of the International Geophysical Year.

A number of authors believe that by the resolution of the International Geophysical Conference, although it was a non-governmental action, a legal basis for the launching of satellites has been established, at the attitude of scientific organizations was supported by the agreement of Governments [6].

The agreement on the IGY undoubtedly represents from the point of view of International Law an important element in the establishment of the legal regime of outer space. Nevertheless, to conclude such an agreement and issue the declarations on the launching of rockets, which — without doing so expressly — also comprised the answer to important legal questions, was just possible because *there does not exist any legal norm forbidding the flights into outer space and in it for peaceful purposes.* The Soviet author A. Galina has correctly deduced that "the non-existence of international legal norms gives reason for the affirmation that any state can freely use the interplanetary space and launch there its satellites and rockets, without asking other states for a permission" [7].

Therefore it is possible to require within the framework of future regime of space the validity for the principle that the *outer space, i.e.* the space located above the airspace, be its frontier later on defined anywhere and anyhow, *is and should remain freely accessible under equal conditions to peaceful research and exploitation by all countries. Similarly to high seas the outer space, too, should be considered as res omnium communis governed by the principle of freedom of navigation which is excluded from the sovereignty of individual states* [8].

III.

The originating legal regime, which would regulate activities of people in the outer space, must grow from the basis of the contemporary international law, from the principles of peaceful coexistence of nations, as these principles regulate the relations between states and their nationals, be they realized on the land and sea, or in the air, or whether they will be realized in the outer space by space vehicles.

The Charter of the United Nations is undoubtedly the most important document in which the principles of peaceful coexistence of all nations are incorporated.

It is possible to affirm that hand in hand with the man's penetration into the outer space also the space validity of the principles of peaceful coexistence, contained in the UN Charter, is extending. The close connection of all spheres, where these principles are applied comes in this connection into the foreground: peaceful coexistence of nations on the Earth must be accompanied by peaceful coexistence in the outer space and vice versa. Without maintaining and ensuring of peaceful coexistence on the Earth, there will not be any certainty on peaceful exploitation of outer space. For this reason the frontier between air space and outer space — whether with or without contiguous zone — could not create any unsurmountable boundary. However paradoxical it may seen, every infringement of state sovereignty in the airspace is simultaneously an attack on the freedom of peaceful flights into outer space, as peace is one and indivisible.

The sovereignty of all states, equal members of international community on the Earth, assumes of course the first place among the principles of peaceful coexistence. Although we express here the principle of delimitation of the sovereignty of individual subjacent states in the space above the state territory up to a certain boundary, we mean by that only its *territorial reach,* which does not mean absolute exclusion of valid principle of sovereignty from outer space in general. The principle of sovereignty radiates into the outer space: freedom of access into the outer space and flights in it, does not mean an unrestricted freedom of action, which would be *e.g.* directed against the integrity and territorial inviolability of states on the Earth. Similarly to the high seas, the freedom of outer space must not be abused for attack on the inviolability of the states or to its endangering; actions aiming at this goal would, therefore justify the states to an appropriate defence and countermeasures [9].

It is necessary to bear in mind this fact especially today, when astronautics greatly develop hand in hand with other technical devices. Today, the far reaching possibilities of practical use of satellites, as wireless re-transmitting stations, for radiotelephonic service, weather forecasting, navigational purposes of air and sea transport, as well as permanent stations of other flights in the outer space *etc.* are clearly outlined. Simultaneously, however, plans and requirements, aiming at the use of these great possibilities especially for war purposes are appearing.

Under these conditions it is necessary to consider simultaneously the requirement of the security of states not only as a problem of the definition of the *boundaries* of the airspace and outer space. The standpoint of security must be also applied to the determination of the *purposes* which astronautics can follow, *i.e. the kind of activities* which may be developed in the outer space in general, and especially in its part, contiguous to the zone of complete and exclusive sovereignty [10].

With regard to the mentioned war-like trends and steadily growing technical possibilities the needs of the security of states could not be properly met by the narrowing of the question only to the height of the respective space column above the state territory. On the other hand, demilitarization and neutralization of outer space would remove many obstacles to the solution of other problems, some of which seem to be unsurmountable under present conditions. It would open the way for example to the delimitation of outer space, favourable to the needs of the future astronautical development.

In this respect the importance of the efforts to achieve a general and complete disarmament which in a number of complex provisions should comprise also complete prohibition and liquidation of all carriers of nuclear arms, including rockets, aircraft and bases of all kind as well as nuclear arms themselves, under an appropriate effective control, is still more outstanding.

References

[1] To rockets, too, of course, fully applies the principle of sovereignty of the subjacent states over the airspace while they are passing trough this space and before they leave it. This viewpoint has been clearly held even by the author of the first treatise on the legal problems of astronautics, the Czech air law expert Dr *Vladimír Mandl*, (who, obviously because he has published his work on astronautics in Germany, is designated once as a "German" author, at other times as an "Austrian" one), in his work *Das Weltraumrecht, ein Problem der Raumfahrt*, p. 18 (Mannheim-Berlin-Leipzig, 1932).

[2] This right point of view has been expressed by the Soviet authors *F. N. Kovalev* and *I. I. Tcheprov, Iskustvenniye sputniki Zemli i mezhdunarodnoye pravo (Sovietsky yezhegodnik mezhdunarodnogo prava*, p. 139, Moscow 1959).

[3] Such proposal was explained by Prof. *Cooper* as early as 1956 at the Annual Meeting of the American Society of International Law and submitted to the Xth ICAO Assembly in Caracas. (Proceedings of the American Society of International Law at its Fiftieth Annual Meeting, Washington, D.C. April 25—28, 1956. Quoted from the "Space Law — A Symposium". Prepared at the request of Honorable London B. Johnson, Chairman, Special Committee on Space and Astronautics, United States Senate, Eighty-Fifth Congress, Second Session, Washington, 1959, pp. 128-129).

Cooper's first point, however, is incorrect. His restrictive opinion, concerning the validity of the Chicago Convention cannot be applied to the principle itself of the sovereignty of states over the airspace, from which the Chicago Convention sets out and which is expressly declared in it *without any restrictions*. For this principle has become a generally recognized principle of international law even before the conclusion of the Chicago Convention (indeed even before the conclusion of the Paris Convention), being an expression of the general principle of the sovereignty of states, on which the whole common international law is based. From the fact that the Chicago Convention regulates civil aviation, which develops in certain, sufficiently dense part of the airspace, it is impossible to infer that it restricts the sovereignty of states only on this part. Regarding the second line, Cooper himself recognized, in a letter to "The Times" of London of Sept. 2, 1957, the necessity of its revision. A serious defect of Cooper's proposal is that it seems to admit an unrestricted freedom of action beyond the frontier of the contiguous zone and outer space, therefore also such action which is aimed against the integrity and inviolability of the States on the Earth.

[4] *Alex Mayer* in a discussion to Cooper's theme (Space Law, p. 134).

[5] *Jaroslav Zourek, Jaky je právní rezim vesmíru?*, Czechoslovak Journal of International Law No 1/1959, p. 42.

[6] E. *Korovine, O mezhdunarodnom rezhime kosmicheskogo prostranstva, Mezhdunarodnaya zhizn* No 1/1959, p. 14. Prof. *Vladimír Outrata*, International Law, Prague 1960, p. 228. *Loftus Becker*, the legal adviser of the State Department, in his statement before the Special Senate Committee on Space and Astronautics on May 14, 1958, (Major Aspects of the Problem of Outer Space, Space Law, p. 37).

[7] *K voprosu o mezhplanetnom pravye, Sovetskoye Gosudarstvo i Pravo*, No 7/1958, p. 55.

[8] See the second of the resolutions accepted to this item at the ILA 48th Conference (International Law Association, Report of the Forty-Eighth Conference held at New York, September 1 to 7, 1958, p. 330).

[9] Therefore it is not also correct to compare the outer space with a kind of "legal vacuum" as *e.g.* expressed by *M. Smirnoff* in the document submitted to the First Colloquium on the Law of Outer Space at the IXth Congress of the IAF in the Hague, 1958, "The Need for a New System of Norms for Space Law and the Danger of Conflict with the Terms of the Chicago Convention".

[10] This substance of the question has been grasped by the Polish author *Jerzy Stucki* in the article *Bezpieczenstwo panstaw a przestrzen kosmiczna, Sprawy miedzynarodowe*, No 7-8/1959, p. 96.

Thoughts on the Importance and Task of Space Law

Franz Gross

> *Give me a fixed point outside the earth,*
> *and I will lift her from her hinges.*
> (Attributed to Archimedes of
> Syracuse, 282—212 B.C.)

The problems of space law cannot be solved just from the earth, *i.e.* by applying earthly standards to the universe, but only from the universe, *i.e.* by the application of new standards both to the universe and to the earth.

Therefore it is not possible to limit either the contents or the area, where space law standards are recognized; for those standards must be the fundamental standards for any other standards in a gradual arrangement of laws.

Thus the sphere of space law standards is not limited either by a limitation of height or by a practical limitation of these standards by astronautics.

The importance of space law standards lies in the fact that they must contain the constitutional standards of the earth, appearing as an entirety against the universe and the celestial bodies.

At present, only international treaties about administration can be made, because we stand only at the beginning of this great human revolution.

I believe that the true significance and thus also the task of space-law is still underestimated. This is also the reason why, for certain of its problems, no satisfactory solution has yet been found. Among these problems must be reckoned, in particular, the type and contents, as also the range, of the regulations to be issued.

When Professor Kelsen, discussing Article II of the Chicago Convention, stated that the sphere of authority and the legal system of any one state is not confined to a surface but exists in three-dimensional space, he did no more than take into account the global shape of our planet.

We must remember, however, that our legislation must from now on take cognizance of the position of our earth in space. We must henceforward, in our legal thinking, make a final effort to progress from Ptolemy to Copernicus.

In other words, the legal problems of space can no longer be properly judged and solved from the starting point of the earth but only from that of space itself. Thus we cannot simply apply terrestrial norms to space but must, on the contrary, see to it that the norms of space are valid on the earth.

This corresponds to the laws of natural and human life as I should now like briefly to sketch them.

We understand by the word "law" the sum total of the norms regulating the human behaviour necessary for the preservation of life; but law is always dependent upon man and upon space. If living-space is restricted, social relations and legal standards become more condensed: if it is extensive, individual freedom increases and private law gains in importance. Thus, in the spatial confines of the ancient and mediæval world men formed "society-states", while in the wide, earth-spanning modern age they formed "surface-states".

In those days there was also born the concept of sovereignty, of which Professor Alfred Verdross says in his work on International Law:

"The concept of territorial sovereignty was formed in conjunction with the Roman conception of property. Both are absolute rights, effective against all and sundry.

> They differ, however, in that territorial sovereignty is a right of disposal under inter-
> national law, while property is a right of disposal on the basis of the international
> law of any one state and is therefore subject to numerous restrictions unknown to
> International Law."

Unknown to International Law *as yet,* because the condensation of society took place at first only within the confines of the surface states, while the condensation of the globe, and with it the juxtaposition of the surface states and the increase in the number of these, has happened only as a result of the development of political and communications factors within the past fifty years.

While at the Paris Conference of 1919 the various nations asserted their absolute property rights in three-dimensional national space by means of exclusive and unlimited air-sovereignty, their citizens had already been restricted for some time in their private right as handed down by the commentators. An evolution had occurred which had led from property and the absolute power of the individual via the limitation of this power by the usufruct rights of all and sundry to its final stage in the restriction of this usufruct by state-power.

A social ordering of air-space had thus already arisen within the various states, while the governments themselves did not as yet recognize this ordering as absolute sovereignty.

Only increasing air-traffic and with it the growing significance of the basic right in International Law to unrestricted world-communications compelled the Chicago Conference of 1944 to concern itself more closely with the concept of exclusive and unrestricted national sovereignty.

Here the concentration and interweaving of the sovereignty claims of the individual states was already visible. The workings of the following evolutionary principle were already becoming apparent: as men thrust forward into new spaces the individual is unconscious of his partnership with other individuals and whith the whole as one limb among many. It is for this reason that the conviction of individual freedom at first prevails. This freedom and individual sovereignty then becomes limited to an ever increasing degree by the freedom and sovereignty of others, while the shape taken by this limiting process forms the basic law of a higher order, namely the order of the whole. This higher order was in the present case national sovereignty in air-space. The state possessed the same unrestricted rights as had been previously enjoyed by the individual.

However, according to the natural principle of entelechy, this process must repeat itself as sovereignty-regulations are merged in a higher order while life thrusts forward into space. That is to say that as nations press upward into outer space their respective sovereignties must be restricted in exactly the same way as individual property rights were once restricted by governments, and this takes place, moreover, according to a social order which corresponds to the position of our earth as a planet in space.

As soon as man leaves the earth he must realize that it is not the centre of, but merely one planet in, the solar system.

Thus space can never assimilate itself to the earth but only the earth to space; which means that our terrestrial order must adapt itself to this higher order.

Clearly it is this fact which Professor E. Korovin also means when he states in his treatise on "The International Status of Outer Space" that space-sovereignty signifies a return from Copernicus to Ptolemy.

The fact, however, that national sovereignty can no longer be exercised in outer space, while on the other hand space works upon national sovereignty, is clearly shown in connection with the problem of neutrality; for in order to preserve its neutrality a national state would have, in the exercise of its defensive sovereignty, to fly over the territories of other neutral states with its defensive weapons and also, as a result of the rotation of the earth, cause belligerent actions above the territories of a number of nations. This would result on the involvement of a neutral state in war with other states by virtue of the very fact that it was defending its neutrality.

From this it follows, however, as Dr. Welf Heinrich Prince of Hanover has several times shown — most recently, I believe, in his contribution to the First Discussion on Space Rights — that the preservation of sovereignty on the earth, if violated from outer space, is virtually impossible.

It is clear, then, that what I might call this "ionisation" of national states must set the same forces in motion as those which our own age has to face as a result of the social revolutions of the past.

As far as the threat of atomic warfare is concerned, it is thus a question of life or death, according to the natural law that life tends continually to expand and must, as it presses forward into new regions of space, either adapt itself to its surroundings or perish.

This fact is also recognized by Professor Oberth in his book "Menschen im Weltraum" ("Men in Space"). He writes:

> "The question of the future of space-travel is at bottom part and parcel of the question of human civilisation as a whole. If this collapses there will be no space-travel either. Research and progress will, to an ever increasing extent, be still possible only when human society is able to do justice to itself, and when all stand together instead of dissipating their strength in quarrels about language, religion, party, system of government, clients and export markets and anxiously trying to keep their knowledge and experience from others."

From the above observations the following conclusions can be drawn respecting the type, contents and sphere of validity of the norms of spatial legislation:

> 1) These must, following the gradations of the legal system, contain basic norms to which our terrestrial norms must adapt themselves.
> 2) In their contents they cannot confine themselves to astronautical traffic but must contain at the same time the constitutional norms of our globe.
> 3) Their sphere of validity is of an all-inclusive nature and cannot be limited to any particular height. It follows
> 4) that we should, in the sense of Mr. Haley's and Professor Cooper's proposals, confine ourselves for the time being to creating international administrative norms, because we are only at the beginning of the greatest revolution in the history of mankind.

The Today Real Possibilities for the Conclusion of an International Convention on Outer Space

Michel Smirnoff

The 1960 was the year in which we waited for a big initiative from the United Nations Organization through its Committee on the Peaceful Uses of Outer Space in the direction of the further steps in organizing the outer space. The *Ad Hoc* Committee created in 1958 was enlarged by six new members as a result of compromise which was reached between the USA and USSR. Nevertheless, 1960 passed without a single meeting of that Committee and, today, it is quite premature to await some important results from an official action on the part of intergovernmental organisms in the way of a codification of future space law.

It is more and more evident that there is an imminent necessity to fill the legal vacuum which does exist concerning the space. More and more nations are prepared to launch satellites and rockets. The danger of still not solved legal problems every day will become bigger and bigger. We cannot wait the formal initiative of governmental and intergovernmental organs. We can only remember the historical facts in the beginning of this century when in 1910 the Paris intergovernmental conference on air law showed that no results could be achieved at that moment. Then the private legal organizations gathered together and formed some Committees to continue the studies of the problems of air law. These studies were the basis on which the Conference of Paris built the first International Convention of Air Navigation of 1919.

The position is now exactly the same. Perhaps the private legal organizations are not the same as in 1910. But there are now many international private legal organizations which are prefectly able to undertake the work which will enourmously help the future effort of intergovernmental organizations in creating space law. There is in the first place the International Astronautical Federation and its International Institute of Space Law which, with its eleven Working Groups, is perfectly able to start on a fruitful effort in preparing the international codification of future space law. There are also the International Law Association with its Committee on Air and Space Law, American Bar Association, French Society of Air Law, and other national and international private legal organizations which have on their agenda the problems of space law.

The time has come when a draft of an agreement or convention on outer space must be prepared as soon as possible, and before the time when the creation of national traditions could make it more difficult to come to an international agreement on this subject. This draft will facilitate the work of a future intergovernmental conference, which will more easily lead to the creation of an International Convention on Outer Space.

Last year in London a paper was presented to the IInd Colloquium on the Law of Outer Space [1], in which was stated that the time had come for the main initiative of private international organizations like ours to begin with serious work in the field of elaboration of rules pertaining the outer space. Today, one year after the

London Colloquium, we can give an account of what happened in respect to official initiative in this field, especially on behalf the United Nations Organization initiative, which the International Astronautical Federation always took as the most important factor in the search for the legal solutions of space problems [2].

We departed last year from London after discussing the very interesting Report of the *Ad Hoc* Committee of the UN on Peaceful Uses of Outer Space [3]. We stated there that this Report was the result of very careful work by the Legal Subcommittee, chairmanned by the well-known Italian lawyer Prof. Antonio Ambrosini. But we also stated that this Report was highly influenced by the political moment and the unusual atmosphere in which the work of this Committee and its Subcommittees developed, with the abstention, not to say boycotting, by 5 members of is initial 18 members. Therefore the highly interesting conclusions of that Report were very cautious and represented a list of priority problems with the expressly stated conclusion of prematurity of the international codification of space law problems [4].

It was quite clear that despite the continued work of the *Ad Hoc* Committee with its limited membership, the reform of that Committee was unavoidable. The Report of the *Ad Hoc* Committee was presented to the Secretary General of UN who had to follow it to the XIV General Assembly of UN. With the mass of problems on the agenda of that General Assembly, discussion of the problem of outer space came only on one of the last days of that session. A resolution was proposed by 12 states, among others USA and USSR, adding another 6 members to the membership of the *Ad Hoc* Committee on the Peaceful Uses of Outer Space, thus putting the number of members at 24. The new members were Austria, Lebanon, Albania, Bulgaria, Hungary, and Rumania. It was quite clear that purely political considerations led to such a Resolution after long negotiations between Henry Cabot Lodge and Mr. Soboleff. The Resolution gave the following tasks to the newly formed *Ad Hoc* Committee:

(a) to work on international collaboration in the outer space;
(b) to find the practical means for the execution of programs for peaceful uses of the outer space;
(c) to create the means for the continuation of the International Geophysical Year;
(d) to organize the exchange of information and to help the national programs of the space exploration;
(e) to continue the studies of the problems of space law;
(f) to prepare the Report to the General Assembly of UN.

The second part of this Resolution speaks about the Soviet proposition to convoke an International Conference on space problems. The Resolution gives a task to the *Ad Hoc* Committee to prepare this Conference, giving the same task to the Secretary General of UN. The Resolution was approved by the General Assembly.

Unfortunately, despite the approval of this Resolution, the Committee in its new formation never met. It is a pity that the reasons were quite insignificant. One of the most important was the problem of Presidency of the Committee for which the Soviet delegation foresaw Mr. Uxa, the Indian Ambassador, and the other members proposed Mrs. Agda Rössel, the Swedish delegate. Another problem was the Presidency of the future World Conference on Outer Space. Before the summit meeting, there was a hope that those difficulties would be put aside, but with the collapse of this meeting the gathering of the Committee is for the moment, according to data available during the writing of this paper, quite uncertain. Therefore, although we are sure that the initiative of UN in this field is the only official way of solving this problem, this initiative for the moment is pretty far from being real and conclusive.

Last year in London we proposed therefore to begin the work within the private international legal organizations, in the first place, within this Federation. Let us see what was done in this respect in the past year.

In London we decided that "the presently constitued Permanent Legal Committee of the IAF be replaced by an International Institute of Space Law, and that an *Ad Hoc* Organizing Committee consisting of five persons and a secretary [5], be authorized to draft bylaws for the organization and government of the proposed Institute, which would be in accordance with the constitution of the IAF, and subject to the approval of the Council of the IAF at a future meeting" [6]. This Resolution was amended by a proposition of well known French lawyer Dr. Eugène Pépin, as follows:

> *"That the General Counsel of the IAF is authorized to establish immediately such Working Groups as are necessary to consider the legal problems of space, which are today considered perhaps capable of resolution, for instance, space radio allocation frequencies, now being considered by the International Telecommunication Union in Geneva, Switzerland"* [7].

Thus in the past year IAF worked on two precise problems and tasks one of which being the creation of the International Institute of Space Law, and the other the establishment of Working Groups for immediate work on some urgent problems of space law.

The first task was carried out by the Organizing Committee headed by Mr. Christopher Shawcross at a meeting of this Committee held in Paris on 20, 21, and 22 March, 1960. Beside Mr. Andrew G. Haley, General Counsel of IAF, and the 5 members of Organizing Committee the meeting was in the presence of Dr. Leslie R. Shepherd, Chairman of British Interplanetary Society, Dr. Frank Malina (USA—UNESCO), George J. Feldman (UN), General Paul Bergeron, President of the French Astronautical Society, and Mr. Gerald Connop Gross, Secretary General of International Telecommunication Union. The main task of that meeting was the discussion of a Project of Statutes of the International Institute of Space Law, prepared by Mr. Andrew G. Haley. The Project of the Statutes had VI articles of which Art. I gave the exact name of the future Institute, Art. II its purposes and objectives, Art. III its membership, Art. IV the meetings of Institute and the functions of its Chairman and Secretary, Art. V the composition of the Executive Committee and its tasks, and Art. VI the details about the members of Executive Committee. After a very fruitful discussion with some remarks, the Statutes were given a definitive form, which had the same number of articles, and its text is now prepared for approval by the Stockholm Congress of IAF.

The name of the future Institute will be "International Institute of Space Law" and the purposes and objectives of the Institute will be:

(a) to provide advice to the President of the Federation when requested;

(b) to carry out such other tasks which may be considered desirable for fostering the social science aspects of astronautics, space travel and exploration;

(c) to publish proceedings and reports and a periodical journal;

(d) to make awards;

(e) to hold meetings and colloquia on juridical and sociological aspects of the social sciences and to make studies and reports;

(f) to adopt, add to, or amend the statutes for the regulation of the internal affairs of the Institute, provided that the Institute shall not enact statutes or amendments thereto which are inconsistent with the provisions of the Constitution of the Federation, or its resolutions pertaining to the Institute.

The members of the Institute will be chosen among the former members of the Permanent Legal Committee of IAF who will in three months time signify their acceptance of membership and of the Statutes of the Institute. The additional members shall be chosen by the *Ad Hoc* Organizing Committee. In the future, the new members will be elected by the Executive Committee of the Institute.

The Institute will hold its annual meetings at the same time as the plenary meetings of the Federation are held. The Institute will have a Chairman and a Secretary. The Executive Committee will be the governing body of the Institute and three or more elected members will constitute a quorum. The main tasks of the Executive Committee will be:

(a) to carry out the purposes and objectives of the Institute.
(b) to implement the resolutions and directives adopted at the annual meetings of the Institute.
(c) to create working groups and committees for all appropriate purposes and functions.
(d) to elect members of the Institute to fill vacancies occuring in the membership of the Executive Committee.
. . .
(g) to arrange for meeting and colloquia.
(h) to arrange for publication of reports and establish a periodical journal.
. . .
(k) to accept donations and legacies, and funds from any private sources, and contributions from national and international nongovernmental and international agencies and from governments.
etc.

Until the first annual meeting of the Institute, the membership of the *Ad Hoc* Organizing committee shall act as the Executive Committee of the Institute. The composition of the Executive Committee will comprise a Chairman, a Secretary and six other members of the Institute to be elected from among the Members which will serve for a period of one year.

Thus, these Statutes will be presented to the Stockholm Congress of IAF in August 1960 for final approval.

The second task was carried out by the initiative of IAF General Counsel Mr. Andrew G. Haley, who was authorized by the amended Resolution of the London Congress of the IAF to create the Working Groups for studies of problems of space law. He sent a circular on January 26, 1960, to the members of the former Permanent Legal Committee with the request to "send . . . a list of ten legal subjects which . . . might be studied by small committees (not more than five members) of the Institute of Space Law". On the basis of this questionnaire Mr. Haley had appointed Working Groups and had assigned the questions for their consideration. The Working Groups were formed from the members of the former Permanent Legal Committee of IAF. There are 11 Working Groups.

The Working Group I, under the chairmanship of the well-known American lawyer John Cobb Cooper, deals with 8 problems from which we mention the delimitation of airspace and outer space and the series of problems connected with the problem of state sovereignty, legal status of space vehicles and outer space.

The Working Group II, under the chairmanship of a great specialist of international law Prof. Alfred Verdross from Austria, discusses the problem of definitions in the space law.

The Working Group III, under the chairmanship of Dr. Michel Smirnoff, will deal with the problem of sovereignty on celestial bodies, legal status of sun and planets, and with the question of property rights in celestial bodies.

The Working Group IV will, under the chairmanship of Italian lawyer Mario Matteucci, deal with the problems of the consequences to the domestic law of new problems of space law.

The Working Group V under the chairmanship of the well-known British lawyer Christopher Shawcross will deal with the regulations of space flights, their registration, traffic rules, safety, *etc.*

The Working Group VI, under the chairmanship of the well-known French lawyer and former Director of the Air and Space Law Institute of McGill University, Montreal, Dr. Eugène Pépin, will deal with the international organizations which will have the authority on the outer space.

The Working Group VII, under the chairmanship of Mr. Andrew G. Haley, former President of IAF and General Counsel of the Federation and one of the founders of space law, will deal with the radio frequencies problem in space and with the question of collaboration with International Telecommunication Union in this field.

The Working Group VIII, under the chairmanship of the well-known French lawyer, Dr. Robert Homburg, will deal with the effects of space activities on private rights, especially in the areas of nationality, citizenship, customs, domicile, crimes, immigration, emigration, ownership of property, torts and contracts.

The Working Group IX, under the chairmanship of the well-known American lawyer Spencer Beresford, will deal with the problem of the responsibility for the damages by the space activities.

The Working Group X will deal with the international organization to be created for the space problems and with the now existing international organizations dealing with space problems (like IAF).

The Working Group XI, under the chairmanship of a Soviet lawyer [8], will deal with the possibilities of conclusion of international agreements on many problems of space law, such as cooperation in space exploration, prohibition of use of artificial satellites and celestial bodies for certain purposes, cooperation in the development of space law, provisions that space problems, not covered by existing law, be settled by negotiation or arbitration, and many other with inclusion of the adoption of an "International Space Navigation Code" analogous to the "International Code of Signals on the High Seas".

As we can see the activities of those Working Groups are covering all the problems of space law, and it is quite clear that the connected activity of all those Working Groups represents a very interesting experience which can promote the official work in the field of the creation of a space law. One thing is sure: this activity will prepare highly interesting scientific material for a future codification of space law on the interstate level.

Although the International Astronautical Federation had a leading role among the private legal international organizations in the field of preparatory works in the field of space law, we cannot forget the activity of other such organizations. In the first place we mention the International Law Association which on its conference in Hamburg from 7—13 August, 1960, has placed, on the Agenda of its

Committee III, the problems of Space Law based on the excellent Report of Dr. D. Goedhuis. Many other scientific organizations, like Congrès International de Fusées et Engins Guidés, American Bar Association, Société Française de Droit Aérien *etc.*, have also on their agenda the problems of space law.

Coming back to the title of our paper we want especially to underline that, today more than ever, the need for a legal regulation of space problems is important on the international level. Late events in world affairs show us that the arguments of space legal problems begin to enter the terrain of international law. The speech of General de Gaulle on the 16th of May, 1960, after the collapse of the summit meeting, where he said that it is very difficult to deal with the arguments of classic notion of aggression when the satellites and other spacecrafts every day, without any permission, circle over the national territories of states with the biggest possibilities to photograph this territory and to send any kind of important reports to their countries. The fact that, on the 1st of June, 1960, a conservative member of British Parliament, Sir Richard Glyn, asked the British Government what kind of measures had it undertaken to carry down the Russian cosmic ship which constantly injures the British space, is another step further on the dangerous way of international conflicts which could be created by the space flights. Although the representative of the British Government answered that there were no such measures prepared, and that this matter was dealt with by a Committee of UN we feel that the moment came when the serious and quick steps had to be taken to prevent in space any possible conflict with heavy consequences. We point out that the rôle of private legal international organizations in this field is very important to help the official interstate organs and the UN in their efforts which are now for some political reasons slowed down. We should like to repeat, like in London 1959, that we consider the IAF one of the most appropriate international organizations to carry out this important task.

References

[1] *M. Smirnoff*, The rôle of IAF in the elaboration of the norms of future space law, paper presented to the IInd Colloquium on Space Law in London on September 4th, 1959, p. 68—75 (Proceedings, Springer Verlag, Vienna, 1960).

[2] The Resolution of the IXth Congress of International Astronautical Federation, Amsterdam, August 1958.

[3] *Eilene Galloway*, The United Nations *Ad Hoc* Committee on the Peaceful Uses of Outer Space, paper presented at the IInd Colloquium on Space Law, in London, September 4th, 1959.

[4] *Edward Wenk Jr.*, Radio frequency control in space telecommunications, pp. 130 and 131 (Washington, 1960).

[5] As members of this Committee were elected six persons in London: Christopher Shawcross (Great Britain; Chairman), Andrew G. Haley (USA), John Cobb Cooper (USA), Dr. Fritz Gerlach (Germany), Robert Homburg (France), and Dr. M. Smirnoff (Yugoslavia).

[6] Minutes of the Plenary Sessions of London Congress of IAF, p. 22.

[7] Minutes of the Plenary Sessions of London Congress of IAF, p. 23.

[8] The General Counsel of IAF could not have the assurance of participation of Soviet lawyers in the works of future Working Groups. Therefore he reserved the places in its Working Groups to the Soviet lawyers without mention their names, but he expressly stated that "it is sincerely hoped that such cooperation will be effected in the very near future".

Draft to an International Covenant for Outer Space — The Treaty of Antarctica as a Prototype

J. Escobar Faria

Article I.

— — —

3. The Contracting Parties which will launch manned spacecraft towards outer space, the Moon, and other planets, may carry the most powerful atomic and ultrasonic devices in order to protect the astronauts against attack from unknown alien races. Such devices shall be used only in self-defense.

— — —

Article III. 1. A new Department in the UN is created by force of this Treaty, named INTERNATIONAL SPACE AGENCY (ISA), under this single jurisdiction of the United Nations General Assembly.
2. The ISA shall have the control of all space activities, shall inspect with priority all outer space programs of the Contracting Parties, and shall be the executor of this Treaty.
3. The ISA shall be constituted by jurists, natural and social scientists, experts, philosophers, educators, and other non-political representatives, in condition of parity, of all Contracting Parties.
4. The ISA shall form its own regulations subject to the provisions of the present Treaty.

Article IV. 1. Any space devices must be identified by the emblems of the launcher nation, being communicated to the ISA before the launching in order to be inspected previously the undertaking.
2. Under the ISA jurisdiction shall be the authorization for the use of orbits to any space devices according to the technical data previously provided by the interested nations, as well as the order to use radio and telemetric bands of transmission from the outer space.
3. All official departments and self-governing agencies of the UN by force of this Treaty shall provide ISA with all necessary information on a priority basis.

Article V. 1. It is within the power of ISA to impose penalties to the infringer signatory nation of this Treaty after resolution of the United Nations General Assembly, and the penalties shall be decided upon by ballot.
2. All penalties shall be of economic order, the infringer nations losing its quota in the international balance of exportation and importation for either a long or a short time according to the grade of the infraction.
3. The ISA shall establish a table of penalties, which will constitute an annex at the end of the present Treaty.

— — —

Article VII. Any nuclear explosion in the outer space, the Moon, and other heavenly bodies, and the disposal there of radioactive waste material shall be prohibited.

Article VIII. The provisions of the present Treaty shall apply to the extra-atmospheric levels of Earth, and from those levels *ad infinitum*.

— — —

Article X. 1. All Member-nations of the United Nations Organization shall constitute the Contracting Parties of the present Treaty.

— — —

On December 1, 1959, after several preparatory sessions, the Treaty of Antarctia was at last signed by the twelve nations which spurred by scientific purposes have shown unequivocal *animus* in the possession and dominum of those frozen lands. The expression 'scientific purposes' is stressed only because it is obvious that a continent comprising inhospitable glaciers and a climate extremely adverse is no place for the normal dwelling of man. Therefore, only scientific purposes could be called upon in order to justify all claims of possession and dominium.

Up to the signature of the Treaty, Antarctica had always been considered as *territorium nullius* therefore susceptible of possession and domination by the first nation able to actually occupy portions of its lands; the possession would be characterised first by an inchoative title and afterwards by acts of sovereignty. Pioneers explorers, including the late Admiral Byrd, raised the flags of their respective countries on parts of those lands, thus establishing inchoative titles. Later on those nations sent expeditions which stayed there for periods of even six months per year, a step that would allow them the right to claim sovereignty. In Brazil, for instance, the principle of *uti possidetis** was applied in 1668, after Brazilian pioneers (known as Bandeirantes) crossed the Tordesillas Line and occupied for Portugal huge and wild areas that should be but were not occupied by Spain. At a later date, when Portugal and Spain were not at war with each other, a new agreement brought about the defeasance of the former divisory line. The Bandeirantes from Brazil did not acquire the lands belonging to Spain by means of *usucaptio;* rather they acted much in the same way as pioneering scientists have done in Antarctica, the main difference being that the South American lands west of the Tordesillas Line belonged to Spain whereas Antarctia had always been nothing more than *res nullius*. The similarity between Antarctica and South America stem from the fact that both the Bandeirantes and the scientists interested in Antarctica established for their respecive countries inchoative titles followed by actual occupation. One might also recall the affair between the United States of America and the Netherlands regarding the possession and dominium of Las Palmas. By means of neutral arbitration the Netherlands was declared the true owner due to the fact that that nation had exercised acts of sovereignty in that island. As to Antarctica (not yet the whole of it) it is to be emphasized that those frozen areas had never belonged to any particular nation, for the continent was *res nullius* up to the very moment some nations flew their flags over there and established themselves in the wastes for periods reaching six consecutive months, thus changing the status of the land.

However it is or was, nothing of what is said above bears any relation with outer space, the Moon, and uninhabited space bodies in a strict sense, but there is a certain similarity in the general situation and that is why I point out the Antarctica Treaty as an excellent prototype. This covenant is a reality and could be used *mutatis mutandis* for outer space. Because the interplanetary space and the heavenly bodies are not linked to our globe or any other planet, and is not interpenetrated with our atmosphere or that of other planets — exactly for this reason — they are simply *res communes omnium,* that is, things that belong to all (including even rational creatures of other civilized' worlds, as I would like to point out).

It is clear now that Antarctica, formerly considered as *territorium nullius,* has now acquired a new status after the covenant was signed; it is now under the principle of *res communis.* For thirty years after the signature of the covenant any one of the signatory nations can exercise inspection of the scientific bases estab-

---

* *Uti possidetis* means *as you now possess,* opposed to *status quo ante,* a diplomatic formula employed in conventions based on territorial possessions of belligerants.

lished in Antarctica by the contracting parties to verify whether or not the terms of the agreement are being respected. Though not disclaiming their possession and dominium there, the twelve contracting parties must obey several rigid clauses related to the peaceful use of Antarctica, one of which prohibits the use of those regions for testing of nuclear devices. This means that the *res communis* principle was considered to be in effect by a large decrease of power of any sovereignty there.

It is my contention that a similar covenant regarding what is named as outer space is of urgent necessity and that Antarctica Treaty is a splendid prototype. That is why I present this paper to the Eleventh Congress of the International Astronautical Federation — the Space Law Colloquium. I stress that it is paramount to present this idea now to the United Nations so that an international agreement may be promoted at once. We space lawyers must proceed hurriedly before human-made chaos reaches the outer space, the Moon, and other uninhabited heavenly bodies with possibly terrible consequences for all of us subjacent living beings, that is to say, mankind.

Also it is necessary to establish at once that no power as yet can claim sovereignty on the Moon for instance (although we know that a Russian device has already reached its surface), because our natural satellite is in fact *res communis* according to jurists and space lawyers all over the world. Thus, then, the Russian device on the Moon has not changed its *status quo,* and no soviet jurist has claimed an inchoative title on behalf of his nation regardless of the epoch-marking achievement.

A covenant for outer space should exhibit identical characteristics with the Antarctica Treaty.

I am most honored to present an extrapolation from the latter, and want furthermore to submit to this conclave an idea that although not a new one is herein presented experimentally drafted for further appreciation. I suppose it is the first time that a draft of this sort is offered in an articulated manner, specially the so-called Space Agency. I name this bureau THE INTERNATIONAL SPACE AGENCY (ISA).

Another idea, this is really new I guess, I want to present, is that our future manned spacecraft *when travelling outside the Earth to the Moon, and other nearby or distant planets,* shall be rigged with all powerful weapons in order to protect our crews against some possible evil intelligences. I stress the fact that outer space, the Moon, and any uninhabited heavenly bodies must be used for peaceful purposes only, but we must, on the other hand, be cautious about our self-preservation. After all our scientific advancement, our astronautical technology, our hard training for space flight, will be useless if we are liable to be destroyed out there by some possible forms of evil intelligence. It is why I am drafting a special paragraph regarding the protection of our future space travels and astronauts.

Draft to an International Covenant for Outer Space
P r e a m b l e

All Nations of the United Nations Organization recognizing that it is in the interest of all mankind that the outer space, the Moon, and all uninhabited space bodies shall continue forever to be used exclusively for peaceful purposes and shall not become the scene or object of international discord;

Acknowledging the substantial contributions to scientific knowledge resulting from international cooperation in scientific investigation in the mentioned alien space and areas;

Convinced that the establishment of a firm foundation for continuation and development of such cooperation on the basis of freedom of scientific investigation there as applied during the International Geophysical Year accords with the interest of science and the progress of all mankind;

Convinced also that a treaty ensuring the use of the mentioned space and areas for peaceful purposes only and the continuance of international harmony over them will further the purposes and principles embodied in the Charter of the United Nations;

Have agreed as follows:

Article I. 1. The outer space, the Moon, and all uninhabited space bodies shall be used for peaceful purposes only. There shall be prohibited, *inter alia*, any measures of a military nature, such as the establishment of military bases and fortifications, the carrying out of military maneuvers, as well as the testing of any type of weapons.

2. The present treaty shall not prevent the use of military personnel or equipment for scientific research or for any other peaceful purpose.

3. The Contracting Parties which will launch manned spacecraft towards outer space, the Moon, and other planets, may carry the most powerful atomic and ultra-sonic rays devices in order to protect the astronauts against attack from unknown alien races. Such devices shall only be used in self-defense.

Article II. 1. Freedom of scientific investigation in the outer space, the Moon, and other uninhabited space bodies and cooperation toward that end, as applied during the International Geophysical Year, shall continue, subject to the provisions of the present Treaty.

3. The International Geophysical Year shall continue in a permanent basis under new name as International Geophysical Exploration (IGE).

3. The IGE shall be under the aegis of the United Nations, and under the jurisdiction of the ISA (*cf.* Art. III. 1.).

Article III. 1. A new Department for the UN is created by force of this Treaty, named INTERNATIONAL SPACE AGENCY (ISA), under the single jurisdiction of the United Nations General Assembly.

2. The ISA shall have the control of all space activities, shall inspect with priority all outer space programs of the Contracting Parties, and shall be the executor of this Treaty.

3. The ISA shall be constituted by jurists, natural and social scientists, experts, philosophers, educators, and other non-political representatives, in condition of parity, of all Contracting Parties.

4. The ISA shall form its own regulations subject to the provisions of the present Treaty.

Article IV. 1. Any space devices must be identified with the emblems of the launcher nation, being communicated to the ISA before the launching in order to be inspected previously the undertaking.

2. Under the ISA jurisdiction shall be the authorization for the use of orbits to any space devices according to the technical data previously provided by the interested nations, as well as the order to use radio and telemetric bands of transmission from the outer space.

3. All official departements and selv-governing agencies of the United Nations by force of this Treaty shall provide ISA with all necessary information on a priority basis.

Article V. 1. It is within the power of ISA to impose penalties to the infringer signatory nation of this Treaty after resolution of the United Nations General Assembly, and the penalties shall be decided upon by ballot.

2. All penalties shall be of economic order, the infringer nation losing its quota in the international balance of exportation and importation for either a long or a short time according to the grade of infraction.

3. The ISA shall establish a table of penalties, which will constitute an annex at the end of the present Treaty.

ArticleVI. 1. In order to promote international cooperation in scientific investigation in the outer space, the Moon, and other uninhabited space bodies the Contracting Parties agree that, to the greatest extent feasible and practicable:
(a) information regarding plans for scientific programs in the mentioned space and bodies shall be exchanged to permit maximum economy and efficiency of operations;
(b) scientific personnel shall be exchanged between expeditions and space stations.
(c) scientific observations and results shall be exchanged and made freely available.

2. In implementing this Article, every encouragement shall be given to the establishment of cooperative working relations with those specialized agencies of the United Nations, and other international organizations having a scientific or technical interest in outer space programs and undertakings.

Article VII. Any nuclear explosion in the outer space, the Moon, and other heavenly bodies, and the disposal there of radioactive waste material shall be prohibited.

Article VIII. The provisions of the present Treaty shall apply to the extra-atmospheric levels of Earth, and from those levels *ad infinitum*.

Article IX. 1. Each Contracting Party shall have the right to designate its Members to ISA in order to carry out any inspection provided by this Treaty.

2. Each Member of ISA shall have complete freedom of access at any time to any plant for space undertakings of the Contracting Parties in order to exercise the inspection.

3. The Members of ISA shall have diplomatic prerogative, but will work each one under subjection of the jurisdiction of their respective States.

4. All Members of ISA shall assemble themselves in New York at the United Nations site inside of two months from the date the present Treaty be signed, and after by intervals in other places in order to exchange mutual information and to consult each other on grounds of common interests with the aims to consider, formulate, and recommend to their respective governments the means to an objective to be fulfilled on the purposes of this Treaty, as follows:
(a) the fostering to all space researches;
(b) the fostering to an international scientific cooperation;
(c) the use of the extra-atmospheric levels of Earth only for peaceful purposes;
(d) the several modalities of the inspection;
(e) the solving of problems related to the excercise of jurisdiction.

Article X. 1. All Member-nations of the United Nations Organization shall constitute the Contracting Parties of the present Treaty.

2. All sovereign States which will be admitted to the UN will further constitute new Contracting Parties of this Treaty, and those already Member-nations which do not want to sign this Covenant will be able to sign it by accession at any time.

Article XI. 1. If any dispute arises between two or more of the Contracting Parties concerning the interpretation or application of the present Treaty, those Contracting Parties shall consult with each other with a view to having the dispute resolved by negotiation, inquiry, mediation, conciliation, arbitration, judicial settlement or other peaceful means of their own choice.

2. Any dispute of this character not so resolved shall, with the consent, in each case, of all parties to the dispute, be referred to the International Court of Justice for settlement; but failure to reach agreement on references to the International Court shall not absolve parties to the dispute from the responsibility of continuing to seek to resolve it by any of the various peaceful means referred to in paragraph 1 of this Article.

Article XII. 1. The present Treaty may be modified or amended at any time by unanimous agreement of the Contracting Parties.

2. This Treaty shall have indefinite time of duration, being forced however to be ratified by all Contracting Parties after 30 years from the moment it became effective.

Article XIII. The present Treaty shall be redacted in all idioms of the signatory States, being the official languages English, French, Russian, and Spanish, which shall be kept on the files of ISA.

Article XIV. This Treaty becomes effective at the date of its signature.

Here is, Gentlemen, how I suppose would run the lines for a draft aiming at an International Agreement for Outer Space, which would likely be able to prevent any man-made chaos over there and the danger for all of us subjacent living beings.

A Research Approach to the Impacts on Society of Peaceful Space Activities

Donald N. Michael

The implications for society inherent in the various on-going and anticipated consequences of space activities require substantial study and research if the salutory opportunities deriving from space activities are to be realized. Examples of consequences which demand study and planning are presented.

Examples are drawn from the problem areas related to

1. the international utilization of weather data and its effects on crop allocation, tourism, and economic competition;
2. the effective utilization of the indoctrination and education capabilities of the communication satellites;
3. manpower training and utilization; and
4. public attitudes toward space activities.

Examples of necessary studies involving economics, sociology, social psychology, and the law are also described.

Session 2
Damage to Third Parties on the Surface Caused by Space Vehicles

Damage to Third Parties on the Surface Caused by Space Vehicles

Introductory lecture

Eugène Pépin

Ladies and Gentlemen:

This morning and this afternoon we have been visiting various parts of the space either at very high, or less high altitudes, and now we should come down through the clouds to consider some more practical questions on the surface of the earth.

Prior to the successful launching of Sputnik I in October 1957 discussions on space law were generally centered on the legal status of space, the upper limit of airspace (*i.e.* atmospheric space), the right of states with respect to space vehicles circulating over their territories. However, Mr Jenks, in his excellent article published on January 1956 in the *International and Comparative Law Quarterly*, mentioned the problem of damages and said, that "it is apparent that rules on the question of damage on earth or to aviation will be necessarily considered, possibly at a relatively early stage". I also mentioned the problem in September 1956 in a lecture before the French Academy "des Sciences Morales et Politiques", and a few days after the launching of Sputnik I, I called the attention of the members of the Canadian Bar on the various categories of damages which may result from the circulation of space vehicles.

A more elaborate paper on the problem of damage, which may be caused by space vehicles was presented, in August 1958 at the First Colloquium on the Law of Outer Space held in the Hague, by Madame de Rode-Verschoor, Professor of Utrecht, with the following title: "Responsibility of the States for Damage Caused by Launched Space Bodies." She submitted three possible solutions:

1. Adoption of the principle of absolute liability;

2. Adoption of the principle of liability for fault; and

3. Establishment of an international guarantee fund for paying damage, caused by satellites (except in the case, where damage is intentionally caused, in which case the state responsible will always have to pay).

Certain today's speakers proposed, as a new solution, one quite similar to the proposal of Madame de Rode-Verschoor.

Since August 1958 a number of papers dealing specifically with damage caused by space vehicles have appeared in legal periodicals. Among these papers, I wish only to mention an article of Mr. Andrew Haley on "Space Vehicle Starts" published in the *University of Detroit Journal* in February 1958, and another of Mr. Spencer Beresford on "Liability for Ground Damage Caused by Space Craft" which appeared in the *Federal Bar Journal of United States;* this later article discusses specially the applicability of the US legislation to these problems. The survey of space law, published at the beginning of 1959 by the Select Committee on Astronautics and Space Exploration of the US House of Representatives, also referred to the problem of liability as an imminent problem.

Furthermore, USSR legal writers are also dealing with the same problem. In an article published in 1958 in the periodical *Soviet State and Law,* Galina included among proposals for an international agreement on space, the liability for injury or damage caused by space vehicles. In January 1959 in *International Life,* Korovin said that that governments should bear full responsibility for personal injury or property loss caused by satellites and other space vehicles. During the same year 1959 the UN legal Subcommittee on Peaceful Uses of Outer Space considered in its report that the problem of liability is one of those which should be studied with priority. The Subcommittee suggested at least five questions:

1. Which kind of damage should give rise to compensation?
2. The nature of acts or negligence involving liability.
3. Should the principles be different when damage is caused on the surface, in atmospheric space or in extra-atmospheric space?
4. Should the liability of the launching state be unlimited?
5. If several states are participating in the launching of space vehicles, should their liability be joint or separate?

The Subcommittee suggested also that in order to secure the payment of any compensation, the states should agree that questions of liability for damage caused by space vehicles should be under the compentency of the International Court of Justice. Reference was also made by the Subcommittee to the Rome Convention of 1952 on Damage Caused by Aircraft to Third Parties on the Surface, which might be taken into account for the consideration of damage to be caused by space vehicles. Up to now we may say that almost all the writers on the question agreed on the necessity to study the problem of damage and some of them add that it is necessary to adopt at an early stage an international convention.

From this very short historical statement, which is obviously incomplete, it appears that up to now no real and general discussion of the problem has taken place among lawyers of the several countries; and I wish to say that the Organizing Committee of this Colloquium has been very wise in giving us today such an opportunity. Several papers have been prepared on the subject and their publication in a volume will certainly be a great contribution for future studies. But it seems to me — and I am sure that I am in agreement with the Chairman — that a comprehensive exchange of views on the most important aspects of the problem in order to arrive to some common views or agreed principles would certainly be a great advantage. Here we are not in a controversial or political problem. There is a common agreement on the necessity to pay certain compensation in case of damage; but it has never been discussed how such compensation is to be paid; how it should be calculated and so forth. Today we have here the possibility to have a more complete exchange of views than before. Such a discussion will not present any difficulty for the distinguished lawyers participating to the Colloquium are quite familiar with the fundamental issues.

I think that we should not complicate the problem. If a damage has been caused by a vehicle of or pertaining to a certain state, in the territory of the same state, this is a national problem; but if we have in a damage elements of different nationalities — that is to say concerning the author of the damage, the victim of the damage, or the territory itself — we have here an international problem on which we should concentrate.

With your permission, I would suggest that we consider successively six important points of that problem, and that some speakers may express on each one their personal views.

Damage to Third Parties on the Surface Caused by Space Vehicles. Discussion

Those points are the following:

1. Should any damage caused by a space vehicle on any part of the surface of the earth give rise to compensation? Or should some distinction be made according to the category of damage, the type of space vehicle, the place of occurence or the victims and so forth?

2. In case of damage, who is liable for compensation? For the present the launching state in question, in the future state- or private organizations or persons, as the case may be?

3. Should the state be considered in any case internationally liable or nationally liable for compensation according to the principle of absolute liability, or should the responsibility be restricted according to the fault principle?

4. Should the liability of the launching state or of the responsible person be unlimited, or limited as regards the amount of compensation?

5. Which jurisdiction, national or international, existing or to be created, may be commonly agreed for consideration of claims arising from such damages?

6. Would it be necessary to conclude a new international convention on this subject? Would it be possible to extend to such claims the application of any existing convention, concerning damages to third parties on the surface?

After Dr. Pépins introduction, a number of papers were scheduled for presentation. It was, however, agreed to go directly into a discussion on the points raised by Dr. Pépin, and to present the papers in the Proceedings, see pp. 141—157.

Discussion

Cooper

Mr. Chairman, Ladies and Gentlemen. I have listened with great care and always with greatest interest and appreciation to my distinguished colleague, Dr. Pépin, whom I had the honour and pleasure recommending to McGill University to succeed me when I retired as Director of the Institute at McGill. I think that my views as to all but one of the questions posed have perhaps been covered in the short paper, which I have filed.

My view is, categorically, that any damage on the surface should give rise to international liability, if the launching state is a state different from the state of the person injured. That will take care the problem of the high seas, that will take care of Antarctica, and I think it is a fairly practical rule.

Secondly, I happen to have been a rather careful student of the recently discussed proposed conventions on catastrophic damage caused by nuclear incidence. Those conventions are not yet in effect. They deal with the problem on the basis of channeling the liability to a single entity to save much trouble. I would say without question, that the liability should be channelled as an international obligation to the state which has launched or authorized the launching of the space vehicle. I think that can be scientifically fairly well followed.

Next. In my paper — and I apologize to the Chairman and Dr. Pépin — I have gone outside of the preview of the stated subject because I have answered the last question first. I came to the conclusion that no adoptation of the Rome Convention or any other convention would be adequate and that a new convention would be required. Therefore I thought that in such a new convention you should deal with:

a. liability on the surface,

b. liability to aircraft and persons therein in the air space, and

c. possible collisions in outer space.

Now may I call your attention to that Echo I, which was launched a few days ago, is about the size of a ten story building. If we have many instrumentalities of that kind operating in space on orbits, which are at 90 degree angle or any orbit except perfectly parallel, we may have very serious collisions in outer space. We are dealing there with an entirely new problem, which is quite definite. May I repeat therefore: my suggestion is that there should be absolute liability of the launching state for damage on the surface, and for damage in the airspace. And then, as to damage in outer space, I believe that the rule should be one of negligence. This is purely a suggestion as we are creating entirely new powers in making that suggestion.

I was quite familiar with the excellent paper referred to by Dr. Pépin in which an international guarantee fund was suggested in order to provide for the payment. I have had some dealing recently with the problem of a suggested international guarantee fund in an entirely different category and they give rise to all kinds of complications. I suggested in my paper something which is quite radical. I have suggested that the state of the person injured should be responsible by a guarantee or otherwise to that injured person up to their own limits, but that the Convention should provide that the state, which has paid out that limited amount to its own national, will have the right to bring a suit in the World Court against the launching state:

a. in recoupment for the amount which it has paid;

b. to collect whatever additional amount is properly chargeable to the loss (I do not know how you can fix a limitation of liability at this time, we do not know enough about it), and

c. that the plaintiff launching state in the World Court should be entitled to collect its own damages if they were in addition.

I can very well see that a state might be injured as well as an individual. I think now that I have answered very shortly the questions proponded by Dr. Pépin. As he knows only too well how he and I suffered with the Rome Convention, I would strongly urge that this problem should be handled by a completely new Convention. I have gone further in my paper however and suggested that this new convention should be ancillary to a convention dealing with the control of outer space.

Verplaetse

I certainly do not want and I am unable to challenge the great authority of Professor Cooper in matters of air law. I will say very briefly, that I agree with the answers he has given to the questions from Professor Pépin.

If you allow me to elaborate and say a few words upon a particular problem, which is the subject of the paper that has been submitted in an abstract to you, I would touch upon the case where the damage would result not exactly from a space activity but would arise out of a conflict of different activities in the air space. Those conflicts of course look askance at sovereignty first and, secondly, they superpose to the horizontal conflicts a regulation in the air space itself.

The fundamental dissimilarity, it seems to me, is precisely in the question of sovereignty. The conflict, in a certain way, has its roots in and is sovereignty. If you have no sovereignty, that means plural power in absolute fashion in a world, there would be no conflicts of laws. Sovereignty has been extended to air space, but is now under pressure of a new technique of which astronautics are altogether the extreme form and the revolutionary intruder. Nevertheless it remains always certain that all activity, whether aerial activity or extra-atmosperic, is going to start on or is going to be guided from our planet and, hence, infected by sovereignty. Therefore, I think in those cases of conflicts between air law and outer space law, we must look first, for analogy, even if we are fully aware that medium and instrumentalities in vertical space are not at all comparable to the age-old impedimenta of the law in horizontal space.

Analogies which we find in horizontal law are of two types. First are conflicts of public law and secondly conflicts of private law. We all know that conflicts of public law could be solved only by an international agreement between sovereigns or by international adjudication. Short of those there was no way out of the conflict. For instance, if a criminal of nationality A has committed a crime in state B and is now in state C, the legal regulation of those three states referring to his behaviour may be different and somehow those states must reach an agreement in order to solve the conflict. That is what has been done by the extradition treaties for instance. In the private law sector international agreements to solve those conflicts are not unusual, but it is not the normal way of clearance. There exists a body of choice of law rules for any case in which there is an extraneous element or, as it has been translated lately in the US by the Director of the Ford Foundation: "a Factor of Foreignness".

Now, let us look to some of those connecting points. The most useful are the territorial connecting points, for instance the *lex rei sitae* or the *lex loci delicti commissi*. Then there are also the personal connecting points: the national law for persons and, in some countries, also for corporate bodies. Nationality applies to ships and planes. These are powerful reasons to apply this connection also to space ships.

I would like to suggest three observations which should guide us (if there is any guidance in this science which should be the science of guidance) in the solution of a conflict between air law and outer space law in matters of damage.

The first is that, to my mind, we should not stick to a single connecting point but that we should, in most cases, look for the preponderant connections and see where the preponderant connections are. There might be several connections, and, in that case, the way out will be to find the preponderant bundle.

Then another point which seems clear to me is that we might perhaps look for the nearest analogy in maritime law, even though, of course, we all know that the high seas are *res communis juris,* whereas the air space belongs to the sovereign domain of the state. What are the cases in maritime law? When there is a conflict in maritime law, what law is applicable? First, for instance, in the case of a collision in territorial waters. Generally the law of the land is then applied. But there is a famous Norwegian case where Norwegian law was applied in a collision of two Norwegian ships in British territorial waters. So here we have not a clear-cut criterion, but we should look for preponderant connections. Because the case was decided in Norway the national law was applied, although the ships had collided in foreign territorial waters. When the collision happens on the high seas I think we must distinguish. If the ships or vessels have the same nationality, we should apply the law of the nationality. That law should also apply, for instance, in a collision between an airplane and a space ship, having the same nationality. When they have not the same nationality, a distinction should be made again, I think, whether there is negligence or not. Generally the law of the non-negligent ship has been applied in the case. When the collision is fortuitous then a fifty-fifty system is suitable. This is my second observation.

My third observation which goes back to the remarks of Professor Pépin is that another connection may enter into the set and this connection is that sometimes, and that is the case in space activity, the substratum of the legal relation is potentially more dangerous on one side of the conflict. For instance, the classical case of an object falling from a military plane onto the earth and killing a peasant, or a small stray tourist plane cutting into an airliner on regular schedule. In those cases the argument of absolute liability enters into the scheme and then it would be reasonable, it seems to me, to apply the law of that part which is not burdened with responsibility.

These are the few words, which I wanted to add with regard to a very particular question, and which, I hope, may be considered at the same time as an answer, *parte qua,* to the questions from Professor Pépin.

Beresford

Mr Chairman. My paper on this subject, I believe, covers all the points raised by Dr. Pépin, although perhaps not in such detail as he may desire. I also find myself in agreement with Professor Cooper's comments, with two exceptions, and I should like to enlarge on those differences.

First, while Professor Cooper agrees that the basis of liability for injury or damage occurring to third persons on the surface should be absolute and without regard to fault, he also suggests that the basis of liability for such injury or damage occurring in outer space itself, between space vehicles for example, should be negligence. I think all I can say is that

I do not understand the reason for this suggestion. It would seem to me that these two kinds of liability should stand on the same footing and that the same considerations of policy and difficulties of proof would apply to both.

Secondly, Professor Cooper suggests that the treaty provisions concerning liability for injury and damage should be ancillary to an agreement on the legal control of outer space. There would be no objection to this, certainly, if we were able to anticipate reaching an agreement on these two subjects at about the same time. But it does in fact appear that we may have considerable difficulty, and it may take some time, for the nations to reach agreement on the first subject, the general legal control of outer space. It would appear to me, therefore, that if we can reach an agreement more quickly on some of these matters, such as liability for injury and damage, we should go ahead and do so, and not wait for the resolution of the larger question with all its policy problems.

von Rauchhaupt

Mr. Chairman. The principal questions have already been dealt with to a certain extent. May I therefore begin at the end and concentrate on the last two questions. This touches also something that Mr. Beresford mentioned. It as a new treaty we need, there is no doubt about it. But when and how?

Anyway, we have got a committee working here in our Institute, that is charged with the preparation of such a kind of treaty. I hope that this committee will be successful. It is difficult to say what is to be put into this treaty. We are no prophets. We cannot say beforehand what may turn up in the meantime. We just know that we need some foundation of the law and that it is to be applied somewhere by some law court or other organization.

This leads me to the next point, Number 5, concerning jurisdiction. Let us enumerate the different possibilities: a national law court, an international law court, an altogether new law court or finally no law court at all. Now, let us see. If we choose a national law court, this will have to decide on questions of international law. We have such an arrangement in the prize court jurisdiction. Prize law is international and is applied by national law courts. There are good chances that such decisions will be quite useful. Sometimes the first decision will not be quite just, but in due time there will be a rectification and then we can rely on the jurisdiction of the prize courts. So, that is a possibility, we leave it open.

Now the second possibility: an international law court. Well, there we got the law court at the Hague, and this is a very fine law court. We must say that of course in the beginning we all had our little doubts, or great doubts about it, but now we are really surprised what good jurisdiction we get there in the Hague. There are fifteen judges taken from all nations of the world, and they all think within the frame of their own culture. Many of them think with their own religion as a background, because their law is part of their religion and cannot be explained without it. Those are the Eastern high religions of the Islam, of the Hindus in India, and of Confucius in China. And we in the West have Christian beliefs; the Old Testament and New Testament form part of our culture. And we use it also in our law, even if we do not say so exactly in our laws, but we follow the rules of our religion. And all these different religions, the European ones and the Eastern ones, meet somehow on a certain equal basis. We call it the morals, and these morals are the same everywhere. If we follow these morals, then we have a good law and a good start for a new treaty.

The third possibility: We might have a special law court for our space law. But as I just suggested, we have got a fine court for international law, and that is the High Court of Justice in the Hague. And this law court, too, is competent for all these questions, which today disturb us so highly. This law court is in fact the only center we have got in public international law. Therefore it seems unnecessary to create something new.

Finally, the arbitration might also be left to the competence of the International Law Court at the Hague.

Concluding Remarks of Session 2

E. Pépin

Mr. Chairman, Ladies and Gentlemen:

I am very pleased that we have been able today to discuss fully the problem, and specially to have the benefit of very constructive proposals made by Professor Cooper, by Dr. Verplaetse, by Mr. Beresfjord, and by Professor von Rauchhaupt. Certainly everyone agree that a new convention is necessary, but the extent or the contents of the convention will depend on further discussions. Reference has been made to the working groups of the Institute of Space Laws which are in charge of preparing a new draft convention. I think these committees will certainly have not only to pick up among its own members some ideas, but to refer to the work made by the other committees of the Institute in order to prepare a comprehensive text.

The principle of the absolute liability should be adopted with certain restrictions, as been said by Mr. Beresford, and I think that this problem is possible to solve. Professor Cooper has perfectly defined what should be the scope of such problems, that is to say that the question should be limited to the case where the victim is of a nationality different from the launching state irrespective of the place of occurrence. But Professor Cooper said we should add to a convention on this topic the control of the space. I think I agree with Mr. Beresford that we should first try to deal with a new convention on damages. All writers on the subject in any part of the world are unanimous on this subject, but I doubt if we are able to reach soon an agreement on the question of control of outer space. A convention on damages, as limited as it may be, might on the other hand possibly be concluded in the near future.

Closing words

Kurt Grönfors

This meeting now comes to an end and I want very heartily to extend my thanks to all the distinguished participants of this meeting and especially to all the speakers who have contributed so well. I think that one result of this meeting is that topic number two is by far more mature now for a final decision than topic number one. Thank you very much. Furthermore, I want to express my gratitude for your giving me the honour to preside today.

The meeting is closed.

Papers   Session 2

Memorandum of Suggestions for an International Convention on Third Party Damage Caused by Space Vehicles

John Cobb Cooper

The United Nations *Ad Hoc* Committee on the Peaceful Uses of Outer Space raised certain questions as to liability for injury caused by space vehicles:

1. For what kind of injuries should recovery be had?

2. What type of conduct should give rise to liability?

3. Should a different principle govern, depending on whether the place of injury is on the surface of the earth, in the airspace, or in outer space?

4. Should liability of the launching State be unlimited, and when more than one State participates, should liability be joint or several?

These problems can best be answered by a multilateral agreement to be entered into by those States who have previously made effective a plan for international control of outer space. This liability convention should cover damage on the surface, in the airspace, and in outer space, based on these principles:

1. Recovery for such injuries as are now specified in the Rome Convention of 1952 for surface aircraft damage;

2. Liability to exist without proof or fault in case of damage on the surface or in airspace; but,

3. Liability for collision or other damage in outer space only on proof of fault;

4. Liability of launching State to be unlimited under subparagraph 2 and limited under subparagraph 3.

Persons suffering damage on the surface or in the airspace would have difficulty to prove source of damage, and to proceed against foreign States or nationals concerned. Therefore, each State should create a guaranty fund to compensate its national victims of space vehicle damage up to an agreed limit, irrespective of nationality of launching State. The State of the national suffering damage should be entitled to compulsory jurisdiction of international court of justice against State or States responsible for launching so as to recover amount already paid to national victims, and also to recover for victims any additional amount due to compensate the damage.

This memorandum is presented as part of the discussion on the topic before the Congress entitled "Damage to Third Parties on the Surface caused by Space Vehicles." It is assumed that the term "space vehicles" includes all flight instrumentalities capable of flight in "outer space," and does not include those instrumentalities now defined as "aircraft" in the Annexes to the Chicago Convention.

The legal section of the 1959 Report of the United Nations *Ad Hoc* Committee on the Peaceful Uses of Outer Space raised, in substance, the following questions as to liability for injury or damage caused by space vehicles:

1. For what kind of injuries should recovery be had?

2. What type of conduct should give rise to liability?

3. Should a different principle govern depending on whether the place of injury is on the surface of the earth, in the airspace, or in outer space?

4. Should liability of the launching State be unlimited, and when more than one State participates, should liability be joint or several?

After thirty-five years of somewhat varied experience with flight law problems. I am convinced that practical uniformity in the field of liability can be accomplished in no other way than by international legislation. Waiting for eventual uniformity through the development of customary rules is impractical and much too slow in so dynamic a field of human action as flight. This does not rule out the value of academic discussion. In fact no other means is so useful in developing the principles on which international legislation may be predicated. It was therefore with the greatest pleasure that I found the present topic on the agenda of the Congress and noted that the discussion was to be introduced by my distinguished colleague, Dr. Pépin. My only concern was that the stated topic was limited to surface damage, which I thought impractical, particularly as the United Nations Committee questions had included damage in the airspace and in outer space.

This memorandum deals with all of the questions raised by the United Nations Committee and suggests their solution by an eventual international convention. No other means can be accepted if useful results are desired. But even this solution should not be sought immediately. It is first necessary that agreement should be reached as to the lawfulness of outer space flight and the international conditions under which it will be permitted. Rules of liability should be determined as part of the international decision and machinery provided for the indemnification of damages suffered as a result of non-aggressive internationally permitted or authorized space vehicle flight. The justification of this proposal is quite simple. Any formal acknowledgement by a State of the far from certain right of other States to engage in non-aggressive outer space flight "over" the territories of such subjacent State without specific authority creates new hazards to the lives and property of nationals.

Somewhat similar considerations have been recognized in the draft international conventions studied by the International Atomic Agency in Vienna, and the OEEC in Paris, dealing with the completely novel legal problems of third party damage suffered as a result of possible commercial nuclear incidents. New national and international responsibilities must be faced. In the area of national legislation, certain States have already adopted legislation under which the State has assumed large parts of the liability for possible nuclear incidents occuring at nationally operated or licensed nuclear establishments.

Similarly, when States eventually agree, as they must, on some type of control or authorization for non-aggressive outer space flight, they should at the same time arrange for the protection of the individual rights of their nationals. This can be done only by an ancillary international liability convention accepted by the same States.

Memorandum of Suggestions for an International Convention

The following principles might be included:

1. Recovery should be provided for damage or loss under such broad categories as are now specified in the 1952 Rome "Convention on Damage Caused by Foreign Aircraft to Third Parties on the Surface". That Convention requires compensation be paid to any person who suffers damage on the surface caused "by an aircraft in flight or by any person or thing falling therefrom". It excludes compensation if the damage results from the mere fact of passage of the aircraft through the airspace in conformity with existing traffic regulations. No logical reason appears why these broad principles could not be adopted to cover damage caused by space vehicles on the surface, in the airspace, or in outer space.

2. As to the type of conduct giving rise to liability, it is suggested that the principles of the Rome Convention should be applied in the case of space vehicle damage on the surface or in the air space. The Rome Convention provides for compensation without proof other than that the damage was caused by an aircraft in flight or persons or things falling therefrom. In other words, no proof of fault is required. While the Rome Convention applies solely to surface damage, it is suggested that the same rule should apply as to damage caused by a space vehicle to aircraft, and to persons or goods thereon, engaged in flight in the air space. Such aircraft will have no better opportunity to avoid a space vehicle falling or passing through air space, or something falling therefrom, than would a person on the surface be able to avoid a falling aircraft. The practical difference in speeds between the space vehicle and the aircraft must be controlling as to the rule of liability.

3. However, as to collisions or other cause of damage in outer space, the rule should be otherwise. Just as the Legal Committee of the International Civil Aviation Organization has indicated, up to the present, that fault should by the basis of an international convention on collision liability between aircraft in flight, so proof of fault should be the rule as to occurences in outer space as between space vehicles.

4. Who should be held liable and for what amount are most difficult questions, particularly when dealing with entirely new problems. This much is certainly true:

a) The launching or control of space vehicles will in practice be a government function for many years to come.

b) Assuming international agreement for a known system of launching control, en route observation, or orbital calculation, it should be possible to determine the government or governments responsible for the launching which eventually results in third party damage.

c) The space vehicle causing damage may be actually operated by an entity other than the government performing or authorizing the launching, and it may be impossible to prove the character or identity of such operator.

These and many other arguments point to the advisability of channeling liability to the government or governments performing or authorizing the launching of the vehicle causing the damage, if ascertainable.

This system of channeling all liability to a single entity has been used in the nuclear liability conventions now under consideration where liability is channeled to the operator of the nuclear establishment responsible for the nuclear material involved in the nuclear incident.

Assuming that the government, or governments, which launched or authorized the launching, are automatically liable, recovery by the person suffering loss may nevertheless be most difficult and slow. It seems doubtful whether launching States will agree both to waive sovereign immunity and to consent to be sued in foreign courts. Also the normal diplomatic processes by which a State may present and pursue a claim for an injured national against the State or States responsible for the launching may be disastrously slow. As indicated earlier in this memorandum, each State which agrees to a future system of authorized non-aggressive outer space flights ought to assume some responsibility to its own citizens for damage on the surface or in the air space resulting from this new hazard. It is suggested that each State should create a guaranty fund or otherwise obligate itself to compensate its nationals suffering such damage from space vehicles up to a fixed minimum amount for each incident. Perhaps the Rome Convention limits should, for uniformity, be applied.

In addition all States which are parties to the convention should agree to be subject to the compulsory jurisdiction of the International Court of Justice so that the State of the national suffering damage on the surface or in the airspace could proceed against the launching State or States, jointly or severally, to recover:

a) by recoupment the amount already paid by the complaining State to its nationals;

b) such additional amount for the benefit of its nationals suffering damage as may be required for full compensation to cover the damage suffered; and

c) such damage as the complaining State itself may have suffered.

It is difficult now to fix any maximum amount to which the liability of the launching State or States ought to be limited. No experience exists on which such a calculation could now be based. Over all and total limitation of liability might be worked out hereafter by a later amendment of the agreement after sufficient time has elapsed to reach reasonable conclusions.

As to claims arising from collision or other damage in outer space, all contracting States should have the right to invoke the compulsory jurisdiction of the International Court of Justice. In proceedings before that Court, any State which has suffered damage, or whose nationals have suffered, could proceed against the launching State or States for full recovery on proof of fault.

The ancillary convention here suggested is in many respects based on presently accepted principles. In other respects it may tread new paths. It is now put forward as perhaps worthy of serious discussion. New problems are before us. They deserve careful study in these very early days of what may be quite novel jurisprudence. International legal experts must not permit themselves to be outstripped too far by their engineering colleagues.

Conflicts of Air- and Outer Space Law

Julian G. Verplaetse

Though the law on the matter of outer space activity will be utterly different from the law in horizontal space and even from that of air space, any realistic approach, according to present experience, starts from the earth.

If used without reference to ultimate validity, some cautious analogy with past human experience is a reasonable starting point, when considering the matter of conflict of laws arising from different activities, aerial and exoatmospheric, in vertical space.

Tentative proposals:

1:o Infrastructure on land: Law applicable: *lex rei sitae* of the aerodrome.

2:o Infrastructure on the High Seas. Law applicable: regulations of the missile area or identification zone.

3:o Airways and Skypaths.

   a) Taking off and landing. Law applicable: law of the missiles or spaceships, on a presumption *juris* and *de jure* of negligence of the aircraft.

   b) Transit of spacecraft through airspace.

      (i) Damage to surface.

      — conflict of laws: rule governing torts with the particular blend of absolute liability.

      — international law: a new over-all convention relating to space activity is advisable but, in view of the shallow success of the Rome Convention 1952 and for other reasons, the prospects are not bright. At any event, coordination with and, perhaps, modification of the Rome Convention would be necessary.

      (ii) Collision.

      Law applicable: law of the non-negligent craft, with a presumption *juris tantum* of negligence of the spacecraft.

I.

It may be superfluous to restate the difference between conflicts of laws in horizontal and in vertical space.

The fundamental *dissimilarity* is that all conflicts in horizontal space are based on sovereignty. Indeed, sovereignty is the essence of the conflict of laws. If there were no sovereignty, *i.e.* plural power in absolute fashion, in our world, there would be no such conflict. Sovereignty has been extended to airspace, but is under pressure of new technique, of which astronautics are altogether the extreme form and the revolutionary intruder.

None the less, it is still true that all activity, whether aerial or exoatmospheric, will start and be guided from our planet and, hence, affected by sovereignty.

Therefore, the first quest is for analogy. True, medium and instrument of activity in vertical space are not at all comparable to the age-old impedimenta of the law in horizontal space. But, since we have not yet a full experience of the new, let's consider some old situations which adumbrate *similarity*.

In golden olden days, when all law was of horizontal compass, we used to distinguish two situations of fact and law.

First, the domestic case, when the whole set of subjects, objects and relationships is confined to one sovereign. This case is unaltered in air- and outer space law. *E.g.* a spaceship launched in one country and causing damage to third parties in the same country would be liable to the law of that country, even if it had, between departure and landing, circled the globe.

Secondly, the case with an extraneous element (factor of foreignness). That case might result in a conflict between different legal systems.

In the public law sector, settlement of such conflicts is a matter for agreement, inclusive international adjudication, or for belligerency. Short of those, the conflict would remain at the stage of opposition [1].

In the private law sector, international agreements to solve conflicts are not unusual, but they are not the normal way of clearance. All legal systems contain some choice of law rules which refer to the law applicable to cases involving an extraneous element.

In matters of air law, some of the conflicts are covered by international agreements. But it should be reminded that the Rome scheme, regarding damage to third parties, has been ratified only by a few States and that further prospect of ratification looks dim [2].

In matters of space law, an overall and comprehensive Convention would be the ideal solution, chiefly with respect to damage to third parties on the surface. But considerable hindrances (insurance, proper jurisdiction and enforcement, *etc*) will obstruct the road to that goal.

At any rate, whether as fundamental principles for guidance in such Conventions or as connecting points in matters of conflict of laws, we must look for, if not rely on, analogy in our search.

Some fundamental connexions are universally used: *lex rei sitae, lex loci delicti commissi,* nationality, *etc.* Nationality has been extended to ships and planes and sometimes to corporate bodies. More even than for ships and planes, powerful reasons will require the granting of a nationality to space ships.

While applying those connexions, three observations should be brought to the fore:

1) To our mind, in most cases, reference to the law applicable is not based on a single connexion, but on preponderant connexions.

2) It is suggested that perhaps the nearest analogy may be found in maritime law, even though the High Seas are *res communis* and airspace belongs to the sovereign domain of the States. An analogy to the point may be the law applicable to collisions of ships [3]:
    a) Collision in territorial waters: law of the land.
    b) Collision on the High Seas:
        (i) Ships of the same nationality: national law.
        (ii) Ships of different nationality:
            A) If negligence: law of the non-negligent ship.
            B) If fortuitous: fifty-fifty.
It might be observed that a new connexion enters into this set: negligence.

3) Again another connexion may interfere when the substratum of the legal regulation is potentially more dangerous on one side of the conflict *e.g.* an object falling from a military plane kills a peasant working on his field or a small stray tourist plane cuts into a big airliner on regular course and schedule. In those cases absolute liability is a solid argument and likely to build the connexion with the law of the other party *i.e.* of the peasant and of the airliner.

II.

The particular case of conflict of laws arising from different activities in vertical space is, on two counts, a novelty in juridical science. Those conflicts not only look defiantly at sovereignty, but they also superpose to the usual horizontal conflict a duality of regulation in the airspace itself [4].

In the summary outline of those conflicts, which follows, outer space law should be understood not as a new law of cosmic extent, but simply as the regulation of space activity insofar as it is actually carried out by human beings of our earth [5].

1) INFRASTRUCTURE ON LAND.

a) Aerodromes and Launching Sites.

It has been proposed that all launching sites should be internationalized and, in that case, adjoining States would have to adapt their legislation on aerodromes, together with ground control and approach, to their international commitments with respect to launching sites. No such internationalization is likely in the near future. Both aerodromes and launching sites being normally under national jurisdiction, the law applicable to conflicts, when they are located in different countries, would be the *lex sitae* of the aerodrome, since clumsy plane activity on the ground would have priority in view of the higher perils of launching.

b) Aircraft and Spacecraft.

Nationality of airplanes, registry—whether national or international—and frequencies of guidance should be kept clearly distinguished from those of outer space activity. This is the only way to avoid conflicts and, perhaps, chaos.

2) INFRASTRUCTURE ON THE HIGH SEAS.

a) Seadromes and Launching Sites—Target Areas.

It is generally admitted that establishment of launching sites and target areas does not infringe upon the freedom of the High Seas. The Soviet Union as well as the United States have established missile target areas over the Oceans. Perhaps that proposition and that practice should not be accepted without reservations [6]. Seadromes are not longer thought of, except as landing rafts for helicopters. Air-carriers or floating ice-stations are not comparable to seadromes.

From first experience it would seem that the law applicable would be that of the State establishing the missile area and not that of the State carrying out air activity. But warning is necessary [7].

An international regulation would be more appropriate but no customary or conventional rule is applicable to the point and no analogy would be workable in the case.

b) Safety Zones.

ADIZ-CADIZ have established specific regulations applicable to aircraft over a large stretch of sea adjoining land, extending sometimes to a distance of 300 nautical miles off-shore [8]. There is no doubt that similar rights would obtain for missile identification [9]. Any such regulation would have priority over air law.

3) AIRWAYS AND SKYPATHS.

a) Take-off and landing paths of airplanes would be subject to the superior rights of rockets and missiles, since the latter could not modify their course. The problem is not so important with respect to possible conflicts or collisions with sky-bound spacecraft, since take-off by the latter is nearly vertical [10]. It is important, however, at the moment of re-entry, inasmuch as the glide-path may trail over several countries and, hence, cross the jurisdiction of their air sovereignty as well as the administrative set-up of their airways. Here the duty would clearly rest with the air regulations of the State overflown, since it may be supposed that, on re-entry, space-craft could not be diverted from its beam. Re-entry being duly authorized, fault would rest *juris et de jure* with air law. Or, the other way round, the law applicable would be that of the spacecraft, since no fault could be shown against it [11].

b) A problem of paramount importance is that of the passage or transit of space-craft through the airspace. Authors, generally, call on analogy of the Law of the Sea and would like to establish, for spacecraft, the right of innocent passage, analo-gous to that enjoyed by ships through the territorial sea [12].

It is submitted that the situation of facts is quite different at sea and in the air:

(i) At sea, the instrument enjoying innocent passage through the territorial waters is the ship. Entering into and egressing from the coastal zones are part of the uniform, non-distinguished shipping activity. In space, the engine crossing the supposedly marginal atmospheric belt is not the proper instru-ment of air navigation but a different artefact, the spacecraft, built to navi-gate in another medium.

(ii) At sea, the territorial belt and the High Sea have the same medium: wa-ter. In space, the air stratum and the outer layers are fundamentally different.

It would be a matter of further experience and study to determine if and how this fundamental difference will exert an influence on the legal regulation. At any rate, innocent passage would not mean free passage. Some such network as ADIZ-CADIZ would have to be worked out. Regulation of the passage of spacecraft would be imperative.

In that regulation two of the most important items to be considered are damage to the surface and collision.

If a spacecraft, while travelling through atmosphere, explodes or plunges back to earth, damage to the surface is not covered by the Rome Conventions of 1933 and 1952. Municipal law or the choice of law rule of the forum would apply. Muni-cipal Law is likely to be based on the rule of absolute liability. The rule of choice of law would be that relating to torts, the *"locus"* being qualified as the place where the damage is actually done. It is an open question whether the Rome scheme should be modified in order to include damage caused by spacecraft from the air or whether a special international instrument should cover that part.

The perfect case of conflict would be a collision between an aircraft and a space-craft in flight (for the purpose: "on regular flight course"). In the absence of any international regulation, liability for any such collision would rest on a presump-tion of fault of the spacecraft. It might be hard to locate the collision—even less its causes—with respect to land and to apply that law. On the other hand, guidance of spacecraft will normally be less flexible than that of aircraft [13]. This, together with the tremendous speed and power of the former, would make it the potentially most dangerous branch of the conflict. Nonetheless, the presumption against the spacecraft would not be *juris et de jure* but *juris tantum* and a fault of the aircraft is not out of the picture.

References

[1] *G. Schwarzenberger:* A Manual of International Law 4th ed. vol. I, p. 111, deems that damage caused by spacecraft in the territory of any other subject of international law is governed by the principle of international responsibility.
This position is unilateral, in view of the fact that space activity is not necessarily on the public law level. Furthermore, conditions of international responsibility may not

be fulfilled, *e.g.* if there is no breach of an international duty or if such breach is unvoluntary, precluding the application of the principle.

See *G. Schwarzenberger: ibid.,* pp. 163 and 168;

M. Sibert: Traité de Droit International Public I, 310;

P. Guggenheim: Traité de Droit International Public, II, 2 (but against the requirement of negligence, p. 49 seq.);

Clyde Eagleton: The Responsibility of States in International Law, 23;

Ch. De Visscher: La Résponsibilité des États, in Bibliotheca Visseriana, vol. II, 89 at 93;

Adolf Schüle in *Strupp-Schlochauer:* Wörterbuch des Völkerrechts, V:o Delikt, Völkerrechtliches, 326, chiefly at pp. 336—337;

Ch.-Ch. Hyde: International Law chiefly as interpreted and applied by the United States, II, 886 seq.;

L. Gould Wesley: An Introduction to International Law, 508;

L. Cavaré: Le Droit International Positif, II, 273.

It should also be pointed out that the application of the principle of international responsibility supposes the exhaustion of local remedies.

[2] For history and status of international conventions on air law see: *Julian G. Verplaetse:* International Law in Vertical Space (1960), 22 seq.

[3] The proposed rules do not state an exclusive solution. *E.g.* the Admiralty has jurisdiction as regards collisions between British ships in foreign inland waters and between foreign ships other than national ships in foreign waters. (*Halsbury's* Laws of England 3rd ed. vol. 1, V:o Admiralty, n. 124, p. 62. *Marsden's* Collisions at Sea, 10th ed. by *K. C. Mc Guffie,* 224—225). As to the law applicable it is always the maritime law as administered in England, though the liability depends, in some cases upon the law of the place where the collision occurs (*Marsden's* Collisions at Sea, 10th ed. by Mc Guffie, 229 and 231).

Moreover, the proposed rules should be considered in the light of the aforementioned general principle of preponderant connexions. See the case Irma Mignon, a collision between two Norwegian ships in English territorial waters, decided by the Norwegian Højesteret according to Norwegian law (*Borum:* Lovkonflikter, 3rd ed., 165).

[4] For the X-15 and Dyna-Soar see the comment of *Andrew G. Haley* on the Report to the National Aeronautics and Space Administration on the Law of Outer Space, compiled by Professors *Leon Lipson* and *Nicholas de B. Katzenbach.*

[5] *Ch. S. Rhyne:* The Legal Horizons of Space Use and Exploration. Address University of South Dakota Law School, 1958, seems to imply that all our existing law is clearly designed in terms of the earth's atmosphere. There is much to be said in favor of this contention, since any human activity, up to now, is permeated by air. Activity outside atmosphere and attraction is a real novelty. *E.g.* what about real property in an environment where a man can carry a five story building on his hands?

[6] The argument—which need not be discussed here—is based on analogous arguments in matters of pollution or fallout. It should be stressed that, as ever, analogy is to be taken *cum grano salis.*

The United States and the United Kingdom have, on various occasions, established danger areas in the Pacific, when testing thermonuclear bombs. Pollution of the airspace and contamination of the sea were the two main arguments alleged against this practice. Proving grounds for missiles would not be so dangerous, unless the use of nuclear warheads give rise to the same uncontrollable pollution. At any rate, there was never a claim that such areas were under the exclusive control of the United States and the United Kingdom. This would be a clear violation of the principle of the freedom of the seas. The only point under discussion was the legality of the establishment of danger areas.

Some authors consider that the establishment of such areas does not infringe upon the freedom of the High Seas. Strange though it may appear for anyone not acquainted with the works of *Grotius,* they may rely on the father of the freedom of the seas, who never advocated a limitless liberty. He considered *i.a.* that sovereignty over a sea area could be acquired by stationing a fleet: *Grotius:* De Jure Belli ac Pacis, (*Kelsey's* translation [1925] vol. II, 214). Other arguments adduced are the legality of naval vessels in formation, manoeuvres or exercises, restrictions in time of war (considering that cold war is not tantamount to peace), and the paramount factor of self defense. See:

Myres Mc Dougal and *Norbert A. Schlei:* The Hydrogen Bomb Tests in Perspective: Lawful Measures for Security in 64 Y.L.J. 1954—55, 649 and the authorities cited;

Myres Mc Dougal and *W. T. Burke:* Crisis in the Law of the Sea: Community Perspectives *v.* National Egoism 67 Y.L.J. 1957—58, 539.

Mc Dougal, challenging the work of the International Law Commission and its absolutistic doctrine of freedom, argues for relativity, expansibility and, above all, reasonableness under community criteria.

See also: UN Conference on the Law of the Sea, Official Records vol. IV: Second Committee, High Seas, General Regime, A/Conf. 13/40, the opinion of the delegates of the United States (p. 15), Canada (p. 20), *etc.*

But there is weighty authority which considers that the Law of Nations relating to the High Seas as well as the law of humanity, and, in particular, the provisions of the United Nations charter and the Trust Agreement, prohibit the establishment of danger areas. See:

Emanuel Margolis: The Hydrogen Bomb Experiments and International Law 64 Y.L.J. 1954—55, 529;

G. Gidel: Les Explosions Nucléaires Expérimentales et la Liberté de la Haute Mer, in Fundamental Problems of International Law: Festschrift für J. Spiropoulos (1957), 191;

Paul de la Pradelle: Les Frontières de l'Air, R.C.A. 1954. II. 117 at p. 147;

Resolution of the UN Conference of the Law of the Sea, April 27, 1958, on the Report of the Second Committee in connexion with art. 2 of the Law of the High Seas, 52 A.J.I.L. 1958, 864, UN Doc. A/Conf 13/I. 56 reprinted in 38 Dep. State Bull. 1111 (1958) and special supplement to the I.C.L.Q. 1958;

Herbert Krüger in *Strupp-Schlochauer,* I, 791, V:o Hohe See;

Declaration of 21 American Republics at Panama in 1939 at the International Conference of American States, 1939—40, A.J.I.L. Supplement 1940, 17.

See also the delegates at the UN Conference on the Law of the Sea, Official Record vol. IV Second Committee, High Seas General Regime, A./Conf. 13/40: Ceylon (p. 14), Romania (p. 16), Bulgaria (p. 19), Czechoslovakia (24 and 45), Ukrainia (p. 31), Japan (p. 44, pointing out that all tests should be prohibited, since even those performed on the territorial sea or on land have an adverse effect on the High Seas).

More cautiously, there is also reluctance to accept A. and H. tests on the High Seas in *Alison Reppy:* The Grotian Doctrine of the Freedom of the Seas reappraised, 19 Fordham Law Rev. Dec. 1950, n:o 3, p. 243 ("act of doubtful validity and, therefore, if to be done, such explosion should only be permitted after due precautions had been taken to protect the rights of other users of the sea", at p. 283).

Shigeru Oda: The Hydrogen Bomb Tests and International Law in Friedenswarte 1956, 126, who puts the question whether the Japanese Government consented to the designation of danger areas before the Bikini tests in 1954. He does not find any such concurrence nor does he consider its possible effect on legality. (It is submitted that such effect could never bear on the status of the High Seas but merely on the contractual relation of the United States and Japan.)

The question of liability arising out of the test is still unsettled. The United States paid Japan a lump sum—far below that requested by the Governement of Japan and the insurance assessment. The sum was paid—and, it would seem, accepted—as *ex gratia,* without reference to the question of legal liability: *Oda:* Art. cit. at p. 127.

[7] As to the analogous cases of the A. and H. test sites, it has never been denied that proper warning was given through the establishment of danger areas. See:

Shigeru Oda: Art cit. at p. 130 seq. For the area of Christmas Islands during the British tests: I.C.L.Q. 1958, 548.

No penalties were provided for entering the danger areas, the penalty being the danger itself. No conflict with air activity seems to have arisen during the tests.

[8] See:

Eugène Pépin: The Law of the Air and the Draft Articles concerning the Law of the Sea adopted by the International Law Commission at its Eight Session Preparatory Document n:o 4 in A/Conf. 13/37 at p. 70.

John Taylor Murchison: The contiguous Air Space Zone in International Law (1955). *Myres S. Mc Dougal* and *Norbert A. Schlei:* The Hydrogen Bomb Tests in Perspective: Lawful Measures for Security 64 Y.L.J. (1954—55), 648 at p. 677 seq.

[9] "Illustration of contiguous zones for security purposes in the missile age is indeed as accessible as the front page of the daily newspaper". *Myres S. Mc Dougal* and *William T. Burke:* Crisis in the Law of the Sea: Community Perspectives *versus* National Egoism, 67 Y.L.J. (1957—58) 539, at p. 583 note 152.

[10] *Cyril E. S. Horsford:* Principles of International Law in Spaceflight. St Louis University Law Journal 1958, 70—78.

[11] Analogy may be found in maritime and railroad transportation.

The rules in matters of collisions between steamer and sailing vessels are to the point. United States and United Kingdom practice hold that a steam vessel must keep out of the way of a sailing vessel, when proceeding on a course involving the risk of collision. The rationale of the rule is that a steam vessel is more perfectly in command and more easily manoeuvered than a sailing vessel. That rule is nearly absolute and applies even when the steamer is encumbered with a heavy or unwieldy tow, though, in that case, negligence of the sailing vessel has sometimes been accounted for.

See the cases cited in *Corpus Juris Secundum* 15 V:o Collision, VI Steam Vessels and Sailing vessels; The English and Empire Digest 41, V:o Shipping and Navigation, Part XII, Collisions, p. 747 (f): Risk of Collision between Steamer and Sailing Vessel. Collision between a train and a motor vehicle shows another analogy. Here presumption of fault rests with the motor car. A train has priority on a fixed track and on private property. It is normally expected to proceed to a time schedule. Of course the engine driver must watch for the signals and take all reasonable steps. But railroad is not like a roadway and the motor car driver knows that the engine driver cannot have that full check that he must be expected to have. See: Lloyds Bank Lt v. British Transport Commission, C. A. Feb. 16, 1956, All E.R. 1956. III. 291; Hazell v. British Transport Commission Q.B.D. Nov. 29, 1957 All E.R. 1958. I. 116; Trznadel v. British Transport Commission, C.A. July 4, 1957, All E.R. 1957. III. 196; Also: Kemshead v. British Transport Commission 1958 I. All E.R. 119, C.A.

Again it should be stressed that analogy is a dangerous instrument. Our case is vastly different from the English steamship colliding on the oceans with an American sailing vessel (The Scotia US Sp. Ct. 1871, 14 Wall 170—189, 20 L. ed. 822) or even from the Dutch plane colliding with the roof of a truck (KLM v. Nederlandse, Antillen, C.A. Netherlands Antilles Jan. 12, 1954, IATA Reports n:o 16, Antillaans Juristenblad 1954, 60).

[12] *Cyril E. S. Horsford:* Art. cit.; *D. Goedhuis:* Draft Report to the International Law Association, 1958, 10.

[13] This paper was delivered when the author was informed of a strong minority scientific thinking, which would give a larger part to human judgement and responsibility in space travel. It is claimed that the astronaut would be able to use his own judgement in steering and should do so in view of the shortcomings of the computer system. See: *Alexander Nyman:* Man or Computer in Space in "Air Force and Space Digest" May, 1960.

It is clear that the terms of that opinion would modify the data on which this paper is based.

Principles of Spacecraft Liability

Spencer M. Beresford

This paper reviews the problems involved in international claims against the owners and operators of spacecraft for personal injury and property damage to third parties occurring on foreign soil.

There is not now any international agreement governing such claims. Yet, in the future, personal injury and property damage on the ground are likely to be caused by spacecraft on a significant scale.

Analogies are drawn with air law and atomic energy law. The possible grounds of liability are examined. The problem of damages is discussed, with special reference to harm resulting from nuclear explosions or radiation. The right to recover and the duty to return strayed spacecraft and their contents under international law are reviewed and related to the question of liability. Finally, consideration is given to the problem of judicial remedies and enforcement.

In conclusion, several provisions are suggested for inclusion in an international agreement on spacecraft liability.

This paper deals with the liability of owners and operators of spacecraft for personal injury and property damage to third parties occuring of foreign soil.

My observations on this subject are personal and not official.

The problem of liability for personal injury and property damage caused by spacecraft is mentioned in almost all writings on space law. As a rule, however, discussion of the subject has been extremely limited [1]. This review of the problems involved in international claims is prompted by the need and prospect of an early governing agreement.

In the future, personal injury and property damage, especially on the ground, are likely to be caused by spacecraft on a significant scale, even without negligence. The times and places at which spacecraft return to the earth will be largely uncontrollable and even unpredictable. Although the risk of ground damage is still slight, it threatens to grow to sizable proportions with the accelerating use of outer space for scientific, military and economic purposes.

A number of missiles and spacecraft have already gone astray. In one instance, ten people lost their lives when missiles exploded on the ground [2].

I. Need for International Agreement

The problem of spacecraft liability is inherently international, since space activities are worldwide by nature and the location of harmful impact will be largely fortuitous.

The unsatisfactory state of international remedies and the diversity of rules under the various municipal legal systems demonstrate the need for establishing uniformity and certainty through an international agreement.

There is not now any agreement governing international claims for personal injury or property damage caused by spacecraft. In the absence of such an agreement, international claims will be burdened with uncertainty and confusion. For example, what laws would govern? What courts would grant a remedy? How could a judgment be enforced?

Air law provides the closest parallel. Appropriate air analogies are the Warsaw and Rome Conventions [3] (which, however, provide for unlike solutions). Another possible analogy is the draft convention on liability for damage resulting from peaceful uses of atomic energy [4].

II. Grounds of Liability

The grounds of spacecraft liability may be viewed differently under various municipal legal systems. Even under Anglo-American law, several kinds of solution are possible: no liability, liability without fault, and liability based on negligence.

It seems not unlikely that spacecraft liability will follow a similar course of historical development to the law of liability for injury and damage caused by aircraft. For some time to come, perhaps, courts may reason that it is not unusual for spacecraft to cause ground damage even though due care is exercised. At first, therefore, the requirements of proof may be very limited, at least for certain kinds and amounts of damage caused by spacecraft. This result may be reached on the theory of absolute liability, or by some other means such as a presumption of negligence or the doctrine of *res ipsa loquitur*. In time, as space technology develops, spacecraft accidents without negligence will become increasingly rare. Finally, when safety requirements and standards of prudent conduct have evolved sufficiently through experience, cases of injury or damage caused by spacecraft may be decided under the general rules of negligence.

Initially, at least, as the foregoing analysis conjectures, the aircraft analogy would suggest the imposition of liability without regard to fault for injury or damage caused by spacecraft.

This conclusion is supported by the fact that the Rome Conventions adopt the rule of liability without fault for ground damage caused by foreign aircraft [5]. Since spacecraft are relatively new and untried, they present a stronger case than aircraft for absolute liability.

The rationale of absolute liability leads to the same conclusion. In Anglo-American law [6] and under the legal systems of many other countries including the Soviet Union [7], liability is imposed without fault for harm resulting from ultrahazardous activities. The rationale appears to be that certain ultrahazardous activities should be permitted because of their social utility, but only on condition that liability is imposed if harm results. Typical examples are blasting, transporting nitrate fertilizers, and, at least until recently, flying airplanes. Similar reasoning could well be applied to spacecraft.

In spacecraft cases, furthermore, proof of negligence is apt to be very difficult. Not only may the necessary evidence be complex and technical, but it may be known only to the Government, and protected by rules of military security.

What has been said above as to the basis of spacecraft liability relates to claims against the owners or operators of spacecraft for personal injury or property damage to third persons on the ground. Different considerations would apply to a claim brought by an ultimate purchaser against the manufacturer of a defective space system (including components and related equipment). Manufacturers of aircraft, for example, have been held liable to their ultimate purchasers [8].

The liability of manufacturers lies beyond the scope of this paper. It is noted, however, that manufacturers' liability in the United States may be based on negligence [9] and limited to personal injury [10].

III. Damages

An international agreement on harm caused by spacecraft must take account of damages. As a practical matter, damages must probably be limited, since States are not likely to assume unlimited liability. On the other hand, the limit of damages should be set high enough to compensate litigants for the greatest injury or loss that can be reasonably expected.

Any limitation of damages for personal injury caused by wrongful conduct may seem to tarnish Anglo-American ideas of tort liability. On reflection, however, parallels come readily to mind. When new rights are created, remedies may be limited. Examples are provided by the "wrongful death" statutes adopted in the various States of the Union. Furthermore, no violence will be done to common-law concepts of damages if liability is imposed without fault.

Adequacy of damages relates to a diversity of national standards. For example, the Warsaw Convention [11] limits damages for personal injury or death to 125 000 gold francs (approximately 8 to 9 thousand dollars). Even though doubled by the Hague Protocol of 1957, this limit is generally viewed in the United States as far too low. Under the Rome Convention of 1952 [12], the limit of damages for personal injury or death is 500 000 gold francs (approximately 30 to 40 thousand dollars). While this amount is still low by American standards, it may well be as much as many countries would accept.

Extensive harm may result from even a single incident involving spacecraft equipped with nuclear engines or warheads [13]. The OEEC draft convention on liability arising from nuclear risks [14] permits recovery of damages up to fifteen million dollars. Even this amount may not be adequate in view of the possible extent of a nuclear catastrophe. For example, the United States Atomic Energy Act authorizes indemnification in any amount up to 500 million dollars for a single "nuclear incident" [15]. In an international agreement on harm caused by spacecraft, it may be necessary to set a separate and higher limit of damages for harm resulting from nuclear explosions or radiation.

IV. Return and Recovery of Spacecraft

The problem of recovery and return of spacecraft (and their personnel, if any) from foreign territory where they have landed by accident, mistake or distress is closely related to the question of liability.

In the absence of international agreement, the right to recover and the duty to return strayed spacecraft and their contents would be governed by the laws of the country in which such spacecraft landed. It is true that off-course civil aircraft and ships are usually returned, but this practice is said to be a matter of comity or specific agreement [16].

Hence the duty to return strayed spacecraft and their contents will remain in doubt unless agreement can be reached. This problem is likely to come to the fore as nations acquire increased reentry and intercept capabilities. The high and growing cost of spacecraft will encourage vigorous efforts for their recovery, especially if passengers or crew members are on board. It seems reasonable to impose a duty to return strayed spacecraft in cases of distress or non-negligent mistake. In addition, the right of a nation to the return or recovery of its spacecraft from foreign soil might well be made conditional upon the acceptance by that nation of liability for injury or damage caused by such spacecraft [17].

V. Judicial Remedies and Enforcement

The International Court of Justice has often been proposed as the proper forum for spacecraft litigation [18]. For example, the United Nations *Ad Hoc* Legal Sub-committee [19] recommended that consideration be given to

> "*agreement on submission to the compulsory jurisdiction of the International Court of Justice in disputes between States as to the liability of States for injury or damage caused by space vehicles*" [20].

In order to be effective, any international agreement on liability for personal injury or property damage caused by spacecraft should provide for the jurisdiction of a court (or courts) and for the enforcement of judgments. Otherwise, the only redress for private citizens, apart from voluntary payment or arbitration, would be the prosecution of their claims through diplomatic channels by their governments. As a result, such claims would founder or, at best, be paid *ex gratia* with no admission of legal fault.

Thus, in general, consideration might well be given to vesting compulsory jurisdiction in the International Court of Justice [21]. But why should the jurisdiction of the court be limited, as the United Nations report implies, to disputes between States? If the jurisdiction is so limited, what remedies will be available to the private individuals and corporations that suffer injury or damage from spacecraft? No matter what forum is chosen, it should entertain private as well as public claims.

Apart from the International Court of Justice, the most likely forum would be the courts of the State where the damage occurred [22]. While this may be the only possible solution, the pitfalls awaiting a national of one State in the courts of another State can be readily imagined.

A third alternative is arbitration, either by a permanent commission or by *ad hoc* commissions composed of representatives from the countries concerned.

Enforcement of spacecraft liability involves other problems than jurisdiction. In particular, sovereign immunity may shut out worthy claims. For some time to come, the only parties likely to be responsible for the flight of spacecraft are national governments, which enjoy some degree of sovereign immunity. Yet in the United States, for example, a citizen damaged by an American space vehicle might not be able to maintain an action under the limited waiver of sovereign immunity granted by the Federal Tort Claims Act [23]. In addition, it may sometimes be difficult to identify the spacecraft that caused the damage. To that extent, dependable judicial remedies may require a system of spacecraft identification and registration.

VI. Conclusions

The considerations described in this paper support inclusion of the following provisions in an international agreement on spacecraft liability:

    a. *Liability without fault for personal injury and property damage to third parties on the ground.*

    b. *Limited but compensatory damages.*
It is suggested that the limit of damages adopted by the Rome Convention of 1952 is the lowest worth considering. For harm resulting from nuclear explosions or radiation, a much higher limit (by a factor of perhaps a thousand) seems appropriate.

    c. *The right to recover and the duty to return strayed spacecraft and their contents, at least in cases of distress or non-negligent mistake, provided the launching State accepts liability for any damage or injury caused by the spacecraft.*

d. *Agreement to the jurisdiction of a suitable tribunal and to the enforcement of its judgments.*

References

[1] For monographs on spacecraft liability, see *Spencer M. Beresford,* Liability for Ground Damage Caused by Spacecraft, 19 Fed.B.J. 242 (1959); *Andrew G. Haley,* Space Vehicle Torts, 36 U.Det.L.Rev. 294 (1959); *I. H. Ph. de Rode-Verschoor,* The Responsibility of the States for the Damage Caused by the Launched Space-Bodies, reprinted in Space Law: A Symposium, Special Senate Committee on Space and Astronautics, 85th Cong., 2nd Sess., pp. 434—5 (1958). See also *Spencer M. Beresford* and *Philip B. Yeager,* Survey of Space Law, Staff Report of the Select Committee on Astronautics and Space Exploration, 85th Cong., 2nd Sess. (1958), reprinted as H.Doc.No. 89, 86th Cong., 1st Sess., pp. 22—26 (1959).

[2] On May 22, 1958, eight Nike-Ajax missiles exploded near Middletown, N.J. One missile launched itself on a two-mile flight. Ten people were killed and three injured. Buildings were damaged for miles around.

[3] Convention for the Unification of Certain Rules Relating to International Transportation by Air (Warsaw, 1929). Convention on Damage Caused by Foreign Aircraft to Third Parties on the Surface (Rome, 1952). More than forty countries, including the United Kingdom, the United States (with reservation), and the USSR, are parties to the Warsaw Convention. The Rome Convention of 1952 has not been signed by either the United States or the USSR, and was signed but not ratified by the United Kingdom. It came into force, however, on February 4, 1958, following the fifth ratification. *Cf.* the Rome Convention for the Unification of Certain Rules Relating to Damage Caused by Aircraft to Third Parties on the Surface (1933).

[4] Prepared for the European Nuclear Agency of the OEEC. See International Problems of Financial Protection Against Nuclear Risk (Atomic Industrial Forum, 1959).

[5] Article 1 (see [3] *supra*). Since contributory negligence is a defense under article 6, it may be more accurate to say that the causing of ground damage creates a conclusive presumption of negligence.

[6] See, for exemple: Rylands v. Fletcher, L.R.3, H.L.330 (1868); Scott v. Shepherd, 2 Black Repts. 892, 96 Eng. Rept. Reprint 525 (1773); Guille v. Swan, 19 Johns 381 (N.Y., 1822). The Restatement, Torts, secs. 519—521 (1938) adopts a somewhat narrower rule of absolute liability.

[7] See Soviet Civil Code, sec. 404.

[8] *E.g.,* Northwest Airlines, Inc. v. Glenn L. Martin Co., 224 F. (2d) 120 (6th Cir. 1955).

[9] *E.g.,* MacPehrson v. Buick Co., 217 N.Y. 392, 111 N.E. 1050 (1916); Smith v. Peerless Glass Co., 295 N.Y. 292, 181 N.E. 576 (1932).

[10] See Restatement, Torts, sec. 395 (1938).

[11] See [3] *supra.* The Warsaw Convention deals with responsibilities of air carriers for passengers and cargo on international flights.

[12] See [3] *supra.*

[13] No known disasters of this kind have occurred to date. On June 7, 1960, however, a Bomarc missile at McGuire Air Force Base, New Jersey, caught fire and released radioactive material. There were no casualties. All personnel on the base were checked for radiation exposure.

[14] See [4] *supra.*

[15] 42 US Code sec. 2210(d).

[16] *E.g.,* Survey of Space Law [1] *supra,* p. 27. See generally *Oliver J. Lissitzyn,* Treatment of Aerial Intruders in Recent Practice and International Law, 47 Am.J.Int.L. 559 ff. (1953). *Contra,* the United Nations *Ad Hoc* Legal Subcommittee on the Peaceful Uses of Outer Space, while suggesting an agreement on the return of spacecraft, considered that

> *"... certain substantive rules of international law already exist concerning rights and duties with respect to aircraft and airmen landing on foreign territory through accident, mistake or duress ...* (and) *might be applied in the event of similar landings of space vehicles."*

See Document A/AC.98/2, United Nations General Assembly, p. 7 (12 June 1959).

[17] See statement of May 7, 1959, to the United Nations *Ad Hoc* Committee on the Peaceful Uses of Outer Space, by Mr. *Loftus Becker,* then Legal Adviser of the Department of State (Press Release No. 3179 of the United States Mission to the United Nations, p. 6).

[18] See, for example, Survey of Space Law, [1] *supra,* at p. 36.

[19] This committee is no longer in existence. An *Ad Hoc* Committee on the Peaceful Uses of Outer Space was established on December 13, 1958, by resolution of the United Nations General Assembly. The report of the Legal Subcommittee (Document A/AC.98/2, United Nations General Assembly) was adopted unanimously by the full committee on June 18, 1959. The Subcommittee consisted of representatives of Argentina, Australia, Belgium, Brazil, Canada, France, Iran, Italy, Japan, Mexico, Sweden, the United Kingdom, and the United States.

[20] See Document A/AC.98/2, United Nations General Assembly, p. 4.

[21] Although the United States has never submitted to the compulsory jurisdiction of the International Court of Justice, there is some evidence that it would favor the Court as a forum for spacecraft litigation. See statement of Mr. Loftus Becker, [17] *supra,* p. 6:

> *"... The real problem in this area is one of enforceability. In this context, therefore, consideration should be given to solving the basic legal problem by recommendation of unqualified submission to the compulsory jurisdiction of the International Court of Justice for any dispute as to a State's liability for injury or damage caused by one of its space vehicles or objects ..."*

[22] For such a provision, see the Rome Convention of 1952, [3] *supra,* article 20.

[23] 60 Stat. 812, 843. See Beresford, [1] *supra,* 19 Fed.B.J. 242, 247—251. Apparently the doctrine of sovereign immunity is less drastic in the civil law.

Appendix

Statutes of the International Institute of Space Law

Approved August 15, 1960, by the 1st Plenary Session of the Delegates to the
XIth International Astronautical Congress, Stockholm 1960.

Article I
Sec. 1. The name of this Institute shall be the International Institute of Space Law
of The International Astronautical Federation.

Article II
Sec. 1. The purposes and objectives of the Institute shall be as follows:
   a. To provide advice to the president of the Federation when requested.
   b. To carry out such other tasks which may be considered desirable for fostering the social science aspects of astronautics, space travel, and exploration.
   c. To publish proceedings and reports and a periodical journal.
   d. To make awards.
   e. To hold meetings and colloquia on juridical and sociological aspects to the social sciences and to make studies and reports.
   f. To adopt, add to, or amend the statutes for the regulation of the internal affairs of the Institute, provided that the Institute shall not enact statutes or amendments thereto which are inconsistent with the provisions of the Constitution of the Federation, or its resolutions pertaining to the Institute.

Article III
Sec. 1. The initial membership of the Institute shall include the members of the
Permanent Legal Committee (which was organized pursuant to the resolution of
the Amsterdam plenary meeting and is now superseded by this Institute) and additional members chosen by the *ad hoc* organizing committee. The initial members
shall signify their acceptance of membership and their agreement to these Statutes
within three months of the approval of these Statutes by depositing a signed letter
of acceptance with the Secretary of the Executive Committee. Thereafter, additional members shall be elected by the Executive Committee of the Institute as
hereinafter provided. All members are elecected for life.
Sec. 2. Applicants for membership in the Institute except the initial members must
be nominated by three Members or by any Member Society. The application shall
be on a form prescribed by the Executive Committee and election shall be by a
majority vote of the members of the Executive Committee in attendance at a
regular or special meeting of said Committee.

Article IV
Sec. 1. The annual meetings of the Institute shall be held each year at such time
and place as the Executive Committee may determine.
Sec. 2. The Chairman and Secretary of the Institute shall also serve, respectively,
as Chairman and Secretary of the Executive Committee.
Sec. 3. The Chairman and Secretary of the Institute and the remaining members
of the Executive Committee shall be elected at the annual meeting of the Institute,
except as provided in Art. VI, Sec. 2.

Article V

Sec. 1. The governing body of the Institute shall be the Executive Committee whose members shall be chosen as provided in Art. VI of these Statutes.

Sec. 2. The Chairman of the Executive Committee, or in his absence or inability to act, the Secretary shall preside at meetings of the Institute and of the Executive Committee. He shall supervise and direct the general business of the Institute pursuant to these Statutes.

Sec. 3. Three or more elected members of the Executive Committee shall constitute a quorum.

Sec. 4. The Executive Committee shall cause minutes to be kept of their meetings and of all action taken by them. Such minutes shall be kept by the Secretary of the Institute and it shall be the privilege of any Member Society of the Federation in good standing to inspect the same at any reasonable time.

Sec. 5. The Executive Committee shall

a. Carry out the purposes and objectives of the Institute as set forth in Article II.

b. Implement the resolutions and directives adopted at the annual meetings of the Institute.

c. Create working groups and committees for all appropriate purposes and functions.

d. Elect members of the Institute to fill vacancies occurring in the membership of the Executive Committee.

e. Supervise the correspondence of the Institute and provide for the safekeeping of the archives thereof.

f. Appoint a clerk, designate his duties and supervise all his activities.

g. Arrange for meetings and colloquia.

h. Arrange for the publication of reports and establish a periodical journal.

i. Recommend concerning the awarding of medals and prizes.

j. Prepare budgets and supervise the auditing of accounts.

k. Accept donations and legacies, and funds from any private sources, and contributions from national and international nongovernmental and international agencies and from governments.

l. Prepare an annual report to be presented at the plenary meeting of the Council of the Federation.

Sec. 6. Until the first annual meeting of the Institute to be held in the calendar year following the approval of these Statutes the membership of the *ad hoc* organizing committee shall act as the Executive Committee of the Institute.

Article VI

Sec. 1. The Executive Committee shall consist of a Chairman, a Secretary and six other Members of the Institute to be elected from among the Members by a majority vote of Members present at the annual meeting of the Institute, beginning with the first meeting after the approval of these Statutes.

Sec. 2. At all times the General Counsel of the Federation shall be an ex-officio voting member of the Executive Committee of the Institute.

Sec. 3. Members of the Executive Committee shall serve until the end of the annual meeting following their election and may be reelected subject to the provisions of Sec. 1 and 2 of his Article.